RIM Fundamentals of Reaction Injection Molding

Christopher W. Macosko

RIM Fundamentals of Reaction Injection Molding

Hanser Publishers, Munich Vienna New York

Distributed in the United States of America by
Oxford University Press, New York
and in Canada by
Oxford University Press, Canada

The Author:
Christopher W. Macosko, University of Minnesota, Department of Chemical Engineering and Materials Science, Minneapolis, MN

Front cover:
Schematic of a RIM machine

Distributed in USA by
Oxford University Press
200 Madison Avenue, New York, N.Y. 10016

Distributed in Canada by
Oxford University Press, Canada
70 Wynford Drive, Don Mills, Ontario M3C IJ9

Distributed in all other countries by
Carl Hanser Verlag
Kolbergerstr. 22
D-8000 München 80

CIP-Titelaufnahme der Deutschen Bibliothek
Macosko, Christopher W.:
RIM, fundamentals of reaction injection molding / Christopher W. Macosko. – Munich; Vienna; New York: Hanser, 1989
    ISBN 3-446-15196-6

ISBN 3-446-15196-6 Carl Hanser Verlag, Munich, Vienna, New York
ISBN 0-19-520759-9 Oxford University Press, New York
Library of Congress Catalogue Card Number 88-083297

To
Kathleen

FOREWORD

The Society of Plastics Engineers (SPE) is pleased to add this volume, "Fundamentals of Reaction Injection Molding," to its ever increasing number of sponsored books. Dr. Christopher W. Macosko is widely recognized as an innovative academician, closely allied with research in new processing developments and their subsequent commercial applications. He is also renowned within the Society as recipient of the 1986 SPE Award in Plastics Research.

This volume represents the maturing of reaction injection molding as a major processing technique. It is written to emphasize fundamental aspects rather than specific technology or potential markets. The presentation skillfully meshes engineering fundamentals with practical aspects of the process. The result is a most understandable discussion of the eight unit operations as well as the materials involved in the technology.

SPE, through the medium of its Technical Volumes Committee, has sponsored books on various aspects of plastics and polymers for over 30 years. Its involvement has ranged from identification of needed volumes to recruitment of authors. An ever-present ingredient has been review of the final manuscript to insure technical accuracy.

This technical competence pervades all SPE activities, not only in publication of books but also in other areas such as technical conferences and educational programs. In addition, the Society publishes five periodicals -- *Plastics Engineering, Polymer Engineering and Science, Polymer Processing and Rheology, Journal of Vinyl Technology* and *Polymer Composites* -- as well as conference proceedings and other selected publications, all of which are subject to similar rigorous technical review procedures.

The resource of some 25,000 practicing plastics engineers has made SPE the largest organization of its type in plastics worldwide. Further information is available from the Society at 14 Fairfield Drive, Brookfield Center, Connecticut 06805, U.S.A.

Robert D. Forger
Executive Director
Society of Plastics Engineers

Technical Volumes Committee
Thomas W. Haas, Chairman
Virginia Commonwealth University

PREFACE

RIM production on a major scale began a little over ten years ago. It is one of the few new polymer processing methods. At first it presented somewhat of a mystery to its practitioners. Superficially RIM looks like TIM (thermoplastic injection molding) but basically it is quite different. We now understand these differences sufficiently well that this book, which focuses on the fundamental aspects of the process, could be written.

About the same time commercial production started, our research group also began to study RIM. This book is largely a review of that work but it also attempts to review all the fundamental studies on RIM. It is intended for those who want to understand how the process works:

- production engineers who must trouble shoot equipment on the plant floor and struggle to optimize the process

- students of polymer processing who need to understand this important and unique process

- chemists who are modifying and designing new polymerizations for RIM

- designers who must determine which process is best suited to make a particular plastic part

- inventors who want to find an even better way!

To get a quick overview of RIM and to determine whether you want to read more, I recommend Chapter 1 followed by the introduction to Chapter 7. The latter is a good summary of Chapters 3 through 6. The key technical concepts of these chapters are summarized in the moldability diagrams discussed in sections 5.4 and 6.4. Properties of major RIM materials are given in Tables 6.3, 7.1 , 8.2 and 8.5. While their formulations are given in Tables 2.4, 7.4, 7.6, 7.7 and 7.8, I hope the subject index will also be helpful to the selective reader.

This book should also be useful to those interested in polyurethane chemistry and structure property relations or in other reactive processes such as composites manufacturing, rubber and thermoset molding. Chapter 2 is an extensive review of how properties develop during polymerization of crosslinked and segmented polyurethanes. It contains over 140 literature citations. Chapter 2 may be passed over on a first reading, especially by those more interested in the RIM process. Important conclusions for RIM from Chapter 2 are summarized in subsequent chapters.

Much of Chapter 6, particularly the conversion and temperature analysis and demolding concepts, are applicable to all types of reactive molding: injection, compression, transfer, potting, encapsulation, autoclave. I believe that Chapter 8.3 is the first review of the fundamentals of resin transfer molding and structural RIM. S-RIM could well be considered another distinct and very new polymer processing method.

Despite how much we have learned about RIM in the past ten years there are still some outstanding problems. At the top of my list are three: 1) better final properties; 2) true automatic production; and 3) computer aided mold design.

Is it possible to reaction injection mold engineering thermoplastics i.e. make high modulus polymers with high use temperature and high impact strength via RIM? Are the high degree of phase separation and high phase connectivity needed for these properties mutually exclusive via RIM?

RIM is still plagued by a high labor content compared to TIM. Further improvement in internal mold release will help but development of flashless molding seems to be essential. Large air bubbles still cause too many repairs or scrap. CAD for RIM molds could help solve this problem. Currently some type of computer assistance is utilized in the design of over half of all TIM parts, but as yet there are no computer aids for RIM. Typically RIM parts are large and thus molds are expensive. Too much time is spent in reworking new molds or operating at less than optimum due to poor design.

If this book helps RIM users avoid previous pitfalls and leads them to new solutions to the problems described above then it will have been worth the effort put into writing it. Criticism and comments from readers are most welcome.

Chris Macosko
Minneapolis
July 1988

ACKNOWLEDGEMENTS

As Lee Blyler of AT&T Bell Labs said upon receiving the Fred O'Connor Engineering/Technology award from the Society of Plastics Engineers, "I would like to thank all the people whose ideas I have borrowed, especially those I forgot to reference, in order to advance my research." I could say the same about my own research and most of the ideas in this book. We all stand on the shoulders of those who went before us.

I would like to thank Frank Critchfield for coming up to me after a seminar in 1973 at Union Carbide in South Charleston. I had spoken about modelling rheological changes during crosslinking of polyurethanes. Frank pointed out the close connections to basic problems in a new process that he and his colleagues (particularly Dick Gerkin, Bill Gill, Lee Lawler, Larry Tackett and later Nigel Barksby) were working on. His early encouragement and the financial support from Union Carbide had a major influence on all our work. The major financial support for this research was provided by the National Science Foundation through the Polymer Program of the Materials Division directed by Norbert Bikales, the Process and Reaction Engineering Program directed by Maria Burka and the former Industry/Cooperative Research University Program.

Additional financial and technical support was provided early in our work by General Motors' Manufacturing Development Division (Joe Hulway, Mike Liedtke, Mary Pannone), ICI (Jim Ferarrini, Herb Gillis, David Spence), American Cyanamid, Monsanto (Jim Gabbert, Mel Hedrick, Dean Kavanaugh, Morris Ort, Marty Wohl), Hercules (Harry Blunt, Steve Ettore, Dick Geer), Martin Sweets (Fritz Schneider), Admiral Equipment and more recently by Dow Chemical (Lou Alberino, Peter Carleton, Marty Cornell, Elio Martinez, Richard Peffley, Bob Turner, Nancy Vespoli), Mobay and Bayer (Ulrich Knipp, Sid Metzger, Georg Schmelzer, Ron Taylor, Hans Wirtz), and Kraus Maffei (Robert Koch and Fritz Schneider). It would have been extremely difficult for an academic to penetrate such a technological subject without all this expert and unselfish advice, and it would be impossible for me to list everyone who helped me to better understand RIM.

One of the greatest pleasures for me at the University of Minnesota is the opportunity to work with such a bright and eager group of students. Barely a word written here would have been possible without them. This book is a small tribute to their tremendous efforts. Although their theses and papers are cited in the text I would like to list the graduate students and post doctoral fellows that I have worked with in the area of RIM. In approximately chronological order they are Fran Mussatti, Steve Lipshitz, Ephriam Broyer, Ed Richter, James Ly Lee, José Castro, Julio Ottino, Paul Kolodziej, Steve Perry, Paul Sibal, Victor Gonzalez-Romero, Libor Matejka, Raphael Camargo, Dave Sandell, Pat Yang, John Blake, Mary Pannone, Ica Manas-Zloczower, Peter Wickert, Zhu Sheng Chen, Kirk Mikkelsen, Thierry Charbonneaux, Steve Machuga, Gibson Batch, Neil Dotson, Wayne Willkomm and Manuel Garcia. Most of them are now doing research or teaching in the polymer field. In addition over 20 undergraduates have helped operate our RIM equipment and done senior research projects in this fertile area. As Jim Lee said after he had built our first mini RIM machine, "I've got a whole chemical plant here." I and my faculty colleagues, Bill Ranz, Ned Thomas, Matt Tirrell and Steve Wellinghoff, have felt that reaction injection molding is an excellent motivator of thesis research. Fundamental questions in diffusion, reaction and phase separation during polymerization, transport calculations, computer aided engineering and the very essence of mixing with reaction all arise when one looks deeply into the process.

This book began as a short course for the Society of Plastics Engineers with Dick Gerkin. It started its transformation from outline to sentences in the winter of 1986 while I was spending a sabbatical year at Institute Charles Sadron, Université Strasbourg, France. Morand Lambla encouraged me to offer a series of lectures on RIM to their graduate students. Tom Haas, publications chair for SPE, urged me on. The rough manuscript was fire polished by students in short courses at Strasbourg, here at Minnesota and with Lou Alberino through the SPE. I enlisted every expert I could for proofreading. Over half of my industrial advisors listed above commented on and corrected parts of the manuscript. In addition Lou Manzione of AT&T Bell Labs read the first five chapters carefully and provided many helpful comments. Gibson Batch, Neil Dotson, Manuel Garcia, Victor Gonzalez-Romero, Sharon Olson and Wayne Willkomm all helped with proofing, making figures and tracking literature citations.

Ed Immergut of Hanser encouraged me to publish and urged me even more to finish it: "A useful book with errors in someone's hands now is considerably more desirable than a perfect one in the indeterminate future." Donna Brogan and Julie Murphy did a masterful job of preparing the camera ready copy.

Tony Ryan from University of Manchester Institute of Science and Technology and currently a NATO postdoctoral fellow here commented extensively on Chapter 2, proofed the final manuscript and helped prepare the subject index.

Certainly I have overlooked others who have helped with this process. My apologies to them and especially to my family and friends who prayed for me and endured like the relatives of someone with an incurable disease. Hallelujah! I am healed! But...let's see, there's still that half-finished manuscript on Rheological Measurements. . .

CONTENTS

1

INTRODUCTION

Reaction injection molding, commonly called RIM, is a method for rapid production of complex plastic parts directly from low viscosity monomers or oligomers. These liquids are combined by impingement mixing just as they enter the mold. Mold pressures are typically very low. Solid polymer forms by crosslinking or phase separation and parts can often be demolded in less than one minute.

RIM is quite different from conventional thermoplastic injection molding (TIM) because it uses polymerization in the mold rather than cooling to form a solid polymer. Other reaction molding processes like monomer casting or thermoset injection molding also use polymerization to set the part shape, however they employ hot mold walls to activate the reaction. In RIM monomer and mold temperature are not so different and the reaction is activated by the impingement mixing.

Figure 1.1 shows a schematic of a RIM machine. Two or more liquid reactants flow at high pressure, typically 100 to 200 bar (1500 to 3000 psi), into a mixing chamber. Usually the flow rate ratio between the two streams must be carefully metered to give the correct stoichiometry of the reactants. In the mixhead the streams impinge at high velocity, mix and begin to polymerize as they flow out into the mold cavity. Because the mixture is initially at a low viscosity, low pressures, less than 10 bar, are needed to fill the mold.

RIM can be broken down into eight unit operations which are illustrated in Figure 1.2. Supply tanks are used to store and blend components. They maintain the level in conditioning tanks at the machine. The conditioning tanks control temperature and degree of dispersion of the reactants by low pressure recirculation. The recirculation loop is also used to inject inert gas (a process called "nucleation" in the industry) which serves to compensate for shrinkage during the mold curing step. The third step in RIM is high pressure metering of the reactants to the mixhead at sufficient flow rate for good mixing and at the proper ratio for complete polymerization. From the impingement chamber the reacting mixture flows into the mold, filling it in typically less than five seconds. There it "cures," polymerizes and solidifies sufficiently to take the stresses of demolding. The final operation consists of various finishing steps including trimming of flash, postcuring, cleaning and

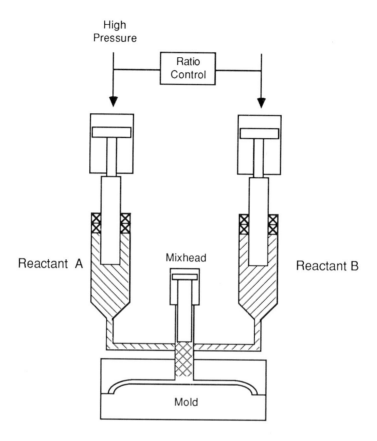

Figure 1.1   Schematic of a RIM machine. When the mixhead ram moves back, two or more liquid reactants flow at high pressure (100-200 bar) into the mixhead chamber. There they impinge and
begin to polymerize as they flow into the mold.

painting. RIM developed from polyurethane rigid foam technology. Since the 1950's urethane elastomers and foam have been made by low pressure injection into flow-through
rotating mixers. High pressure impingement mixing to produce urethane foams was first
reported by Harreis (1969). Pahl and Schlüter (1971) describe the first RIM equipment
with a self cleaning, recirculation mixhead. These machines were first used in Germany to
produce integral-skin, rigid urethane foam for automotive and furniture (Wirtz 1966, 1969;
Piechota, 1970). Major growth of RIM was pushed by needs in the U.S. automotive mar-

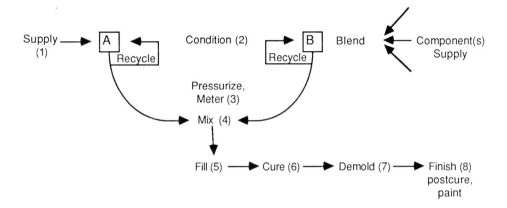

Figure 1.2   The eight unit operations for the reaction injection molding (RIM).

ket. In 1972 Congress mandated that bumpers on all cars sold in the U.S. withstand a 5 mile/hour impact without damage. Flexible polyurethane fascia on the front and rear of cars covering a steel beam mounted on shock absorbers were found to meet this requirement. RIM quickly proved the most economical way to make these large parts and production began in 1974. RIM production in the U.S. grew from 2000 tons in 1974 to 17,000 in 1978 (Becker, 1979). In 1979 glass reinforced urethane products appeared and were applied to automotive fenders and body panels.

By 1987 North American production rose over 70,000 tons with about 600 RIM machines in use. It was still concentrated primarily in exterior automotive parts, but there was significant application for interior auto parts, furniture, business machine housings, construction items such as window frames, appliances and recreational equipment (Poole, 1985). RIM markets for 1987 are shown in Table 1.1. Recent growth has been about 13% per year and is expected to continue. European and Asian production combined is about the size of that in North America. However, in Europe, fascia production is much lower but there are more furniture and appliance applications.

As indicated in Table 1.1 over 95% of RIM production is in polyurethanes or urea-urethanes. Nylon 6 RIM production began in 1983 and dicyclopentadiene and acrylamate systems were introduced in 1985. Development work has been reported on epoxies, unsaturated polyesters and phenolic RIM materials. These different RIM chemical systems are summarized in Table 1.2.

Table 1.1 North American RIM Applications for 1987[a]

	10^3 tons
Auto, fascia	51
Auto, window gaskets	5.5

Other auto
  - fenders, panels
  - exterior trim
  - window gaskets
  - arm rests
  - steering wheels
	~5

Recreation
  - snowmobiles
  - ski boots
  - golf carts
	2.6

Agriculture - vehicle panels	1.3
Furniture	1.0

Electronic
  - business machine housings
  - frames
	0.7

Oil field	0.7

Other
  - appliances
  - housing construction
    (window units)
  - beer keg ends
	5

	72

[a] Poole (1985); Metzger (1987); Cornell (1987) over 95% polyurethane or urea-urethane (*Plastics Techn.* Mar 1986, pg 93).

Table 1.2   RIM Chemical Systems

Polymerization Reaction	T_o °C	T_{mold} °C	$\Delta T^{(a)}$ exotherm	t_{demold} s
urethane				
$-NCO + HO- \xrightarrow{Sn} -\overset{\overset{\displaystyle O}{\|\|}}{\underset{\underset{\displaystyle H}{\|}}{N}}CO-$	40	70	130	45
urea				
$-NCO + H_2N- \longrightarrow -\overset{\overset{\displaystyle O}{\|\|}}{\underset{\underset{\displaystyle H}{\|}}{N}}C\underset{\underset{\displaystyle H}{\|}}{N}-$	40	60	110	30
nylon 6				
$(CH_2)_5-NH, C=O \xrightarrow[Acrylactam]{M_gBr} \left(-(CH_2)_5-\overset{\overset{\displaystyle O}{\|\|}}{\underset{\underset{\displaystyle H}{\|}}{N}}C-\right)_n$	90	135	40	90
dicyclopentadiene				
$\xrightarrow[W Cl_6]{AlEt_2Cl}$	35	60	170	30
polyester				
$CH_2=CH(\phi O) + $ unsaturated esters $\xrightarrow{R \leftarrow \bullet}$ crosslinked network	25	120	~100	60
acrylamate				
$-OH + OCN- \xrightarrow{Sn}$ unsaturated urethane; unsaturated urethane $\xrightarrow{R\bullet}$ crosslinked network	25	100	~100	60
epoxy				
$-\overset{}{CH}\overset{}{-}CH_2 (O) + H_2N \xrightarrow{Cat} -\underset{\underset{\displaystyle OH}{\|}}{CH} CH_2\underset{\underset{\displaystyle H}{\|}}{N}-$	60	120	~150	90

a) adiabatic value for a typical formulation without filler

The major reason for the growth of reaction injection molding is low viscosity during mold filling. Starting with low viscosity liquids a relatively small metering machine can make large parts; 50 kg pieces have been produced (Sneller, 1983). Complex shapes with multiple inserts can readily be fabricated. Low viscosity and low pressures during filling translate into lighter weight and lower cost molds. This opens up short production runs and even prototype applications to RIM. It should be noted, however, that for high production applications like automotive fascia durability requirements result in mold costs similar to those for thermoplastic injection molding (TIM).

Low viscosity also opens many options in reinforcement. One is to place a long fiber mat into the mold before injection. Thus RIM can be used for high speed resin transfer molding (Gonzalez and Macosko, 1983; Eckler and Wilkinson, 1986; Carleton, Waszeciak and Alberino, 1986).

In RIM the mold is packed by gas expansion rather than by high pressure as with TIM. This means that mold clamps can be much smaller and less expensive especially for large parts (Scrivo, 1984). Also, as Table 1.2 indicates, RIM temperatures are typically lower than those for TIM. This means lower energy costs. The important differences between RIM and TIM are summarized in Table 1.3.

Table 1.3   Comparison Between Typical RIM and Thermoplastic Injection Molding[a]

	RIM	TIM
Temperature		
reactants	40°C	200°C
mold	70°C	25°C
Material Viscosity	0.1 - 1 Pa·s	10^2 - 10^5 Pa·s
Injection Pressure	100 bar	1000 bar
Clamping force (for 1 m² surface part)	50 ton	3000 ton

(a) from Lee (1980)

Special materials are also possible with RIM since the polymer is formed directly in the mold. For example, the hydrogen bonding in polyureas is so strong that they do not melt flow like normal thermoplastics. Thus processing by TIM is impossible, yet excellent polymers are produced via RIM (Dominguez, 1984). Polydicyclopentadiene is a cross-linked polymer which must be formed in the absence of oxygen and water (Geer, 1983). RIM appears to be the only practical way to produce it.

But some of the disadvantages of RIM also arise from its low viscosity. For example gas bubbles can become trapped during filling. It is difficult to seal molds and the resulting flash can increase labor costs. Handling of reactive and often hazardous liquids requires special equipment and procedures. Mold release has been a major problem for high production RIM. In general low viscosity liquids penetrate mold surfaces and urethanes, in particular, adhere well to metals. Typically a manual spray of external mold release has been required after every injection. The recent introduction of internal mold release has reduced this to one spray for every 20 to 100 moldings (Cekoric, Taylor and Barrickman, 1983; Meyer, 1985). Such developments along with robot demolding and trimming should help to achieve highly automated RIM production (Lloyd and Cornell, 1985; Sneller, 1986).

The purpose of this monograph is not to discuss market potential or RIM technology in great detail but rather to emphasize the fundamental aspects of reaction injection molding. There are a number of descriptive reviews available. Becker (1979), Sweeney (1987) and Oertel (1985) give RIM machine schematics, mold design information, urethane formulations, process cost analysis and equipment layout ideas. Kresta (1983, 1985) has also compiled information in these areas. Some fundamental concepts in impingement mixing, curing kinetics and urethane structure-property relations were reviewed by Lee (1980). Coates and Johnson (1981) discuss basic aspects of reinforced RIM. Allport, Barker and Chapman (1982) review some polyurethane RIM and reinforced RIM formulations and properties. However, a comprehensive and up-to-date treatment of RIM fundamentals is needed. Our research group at the University of Minnesota has studied reaction injection molding for the past 10 years. This monograph is drawn primarily from that work but includes considerable outside literature in an attempt to develop an understanding of RIM in terms of engineering science. It had its beginnings in a short review article (Macosko, 1983) and hopefully it will provide a basis for continued growth of RIM into new chemistry and new applications.

RIM began with polyurethanes and as Table 1.2 indicates they, along with polyureas, still constitute over 95% of RIM production. Since RIM equipment has evolved around urethanes and ureas, we will focus primarily on them. Thus the next chapter describes urethane chemistry, reaction kinetics rheology and other property development with an emphasis on those aspects which control the RIM process. Chapter 7 describes the other polymerization systems used for RIM, while Chapter 8 treats special problems in reinforced RIM.

Chapter 3 describes RIM equipment. No attempt is made to list all the different manufacturers and the variations between them but focus on the key components of the equipment, particularly the metering operation. The remaining chapters will be divided according to the unit operations shown in Figure 1.2: mixing (Chapter 4), filling (Chapter 5) and curing, demolding and finishing operations (Chapter 6). Because Chapter 7 introduces other chemistry into the RIM process, it serves as a good overview of Chapter 3 through 6.

REFERENCES

Allport, D.C.; Barker, C.; Chapman, J.F. "Developments in polyurethane block copolymer systems for reaction injection molding," in "Developments in Block Copolymers-1," Goodman, I., ed., Applied Science: London, 1982.

Becker, W.E., Ed. "Reaction Injection Molding," Van Nostrand Reinhold: New York, 1979.

Carleton, C.S.; Waszeciak, D.P.; Alberino, L.M. "A RIM process for reinforced plastics," Session 9-D, Proc. of Reinforced Plastics/Composites Inst., Soc. Plast. Ind., Atlanta, January, 1986.

Cekoric, M.E.; Taylor, R.P.; Barrickman, C.E. "Internal mold release, the next step forward in RIM," SAE Tech. Paper #830488, Detroit, March 1983.

Coates, P.D.; Johnson, A.F. "An introduction to reinforced reaction injection molding," *Plast. and Rubber Processsing and Applic.* **1981**, *1*, 223-238.

Cornell, M.C., "Trends in automotive RIM chemistry," Proced. of Polyurethanes World Congress, Soc. Plast. Ind., Aachen, 1987, 198-204.

Dominguez, R.J.G. "Amine-terminated polyether resins in RIM," *J. Cellular Plast.* **1984**, *20*, 433-436.

Eckler, J.H.; Wilkinson, T.C., "Composite manufacture using the reaction injection molding process," Session 9-B, Proc. of Reinforced Plastics/Composites Inst., Soc. Plast. Ind., Atlanta, Jan. 1986.

Geer, R.P. "Polydicyclopentadiene: a new RIM thermoset," Proc. Nat. Tech. Conf., Soc. Plast. Eng., Detroit, 1983, 104-105.

Gonzalez, V.M.; Macosko, C.W. "Properties of mat reinforced reaction injection molded materials," *Polym. Comp.* **1983**, *4*, 190-195.

Harreis, J. "Reactionsgussein neues Verfahren zum Herstellen Grosser Kunststoff-Formteile," *Kunststoffe* **1969**, *59*, 398-402.

Kresta, J.E., Ed. "Reaction Injection Molding and Fast Polymerization Reactions," Proc. of Int. Symp. on Reaction Injection Molding, Am. Chem. Soc., Atlanta, March, 1981, Plenum: New York, 1982. Kresta, J.E., ed. "Reaction Injection Molding," Am. Chem. Soc. Symp. Series 270, Washington D.C., 1985.

Lee, L.J. "Polyurethane reaction injection molding: process, materials and properties," *Rubber Chem. Tech.* **1980**, *53*, 542-599.

Lloyd, E.T.; Cornell, M.C. "Polyurethane RIM: a competitive plastic molding process," in "Reaction Injection Molding," Kresta, J.E., Ed., Am. Chem. Soc. Symp. Series 270, Washington, D.C., 1985.

Macosko, C.W. "Insights into molding RIM materials," *Plast. Eng.* **1983**, *39*, 21-25.

Metzger, S.H., Mobay, personal communication, 1987.

Meyer, L.W. "Self-releasing urethane molding systems: productivity study," in "Reaction Injection Molding," Kresta, J.E., Ed., Am. Chem. Soc. Symp. Series 270, Washington, D.C., 1985.

Mooney, P.J. "RIM: hype or hope?" report P-054R, Business Communication Co: Stanford, CT, June 1983.

Oertel, G., Ed. "Polyurethane Handbook," Hanser: Munich, 1985.

Pahl, F.W.; Schlüter, K. "Processing fundamentals for molded foam polyurethanes I. Storage, metering and mixing of the components," *Kunststoffe* **1971**, *61*, 540-544.

Piechota, H. "A new polyurethane structural foam," *Kunststoffe* **1970**, *60*, 7-14.

Poole, A. "RIM polyurethane use to reach 186 million pounds by 1989," *Plast. Eng.* **1985**, *41*, 12.

Scrivo, J.V. "It pays to mold large parts from RIM structural foam," *Plast. Eng.* **1984**, *40 no. 3*, 67-72.

Sneller, J.A. "RIM for the big parts," *Modern Plast.* **1983**, *60 no. 3*, 24-25.

Sneller, J.A. "Detroit's plastics planning sparks rise in RIM automation," *Modern Plast.* **1986**, *63 no. 2*, 55-58.

Sweeney, F.M. " Reaction Injection Molding, Machinery and Processes," Marcel Dekker: New York, 1987.

Wirtz, H. "Semi-rigid urethane foams in the automotive industry," *J. Cellular Plast.* **1966**, *2*, 324-330.

Wirtz, H. "Integral skin urethane foam molding," *J. Cellular Plast.* **1969**, *5*, 304-309.

2

POLYURETHANES

RIM is an example of reactive polymer processing: polymerization occurs simultaneously with setting of the polymer shape. In all reactive molding processes the parts are
removed when hot. Solidification is due to polymerization in the mold cavity rather than
heat transfer as in thermoplastic molding. The solid structure must result from either
crosslinking or phase separation. This causes the rheological changes shown in Figure
2.1. Initially the viscosity must be low to allow rapid mixing. Then viscosity builds up as
the monomers react. This build up must be fast but not so rapid as to prevent filling large
molds. The modulus and tear strength typically develop more slowly. They limit demolding of the part and thus productivity.

The goal of this chapter is to understand these rheological changes, how they result
from the polymerization kinetics and the buildup of molecular structure. Since polyurethanes and polyureas constitute the vast majority of all RIM production, it is more important to study their chemistry and properties. Polyurethanes can build a solid structure in
the mold by either crosslinking or phase separation. Thus we can use them to illustrate the
concepts which are generally applicable to other RIM chemistry, such as polyamides and
polydicyclopentadiene, treated in Chapter 7.

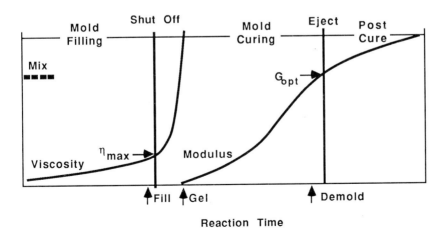

Figure 2.1   Rheological changes during a RIM cycle (adapted from Broyer and Macosko, 1976).

2.1 Reactions

Urethane chemistry is the chemistry of the isocyanate group. Thus polyurethanes and polyureas are usually treated together. Isocyanates react readily with compounds containing an active hydrogen, i.e. a hydrogen which is replaceable by sodium. The active hydrogen compound adds across the carbon-nitrogen double bond of the isocyanate group.

$$R\text{-}NCO + HA \rightarrow R\text{-}NH\text{-}CO\text{-}A \qquad (2.1)$$

For polymerization the most important active hydrogen compounds are alcohols which react to form urethanes

$$R\text{-}NCO + HO\text{-}R' \rightarrow R\text{-}NH\text{-}CO\text{-}O\text{-}R' \qquad (2.2)$$

and amines which form ureas.

$$R\text{-}NCO + H_2N\text{-}R' \rightarrow R\text{-}NH\text{-}CO\text{-}NH\text{-}R' \qquad (2.3)$$

Isocyanates can react with urethanes to form allophanates and with ureas to form biurets. These reactions result in crosslinking, but, as Table 2.1 indicates, their rates are at least 10^2 less rapid than those needed for RIM.

Water reacts with isocyanate to form a carbamic acid which in turn can break down into an amine and CO_2. The amine then reacts with another isocyanate to form a urea

$$R\text{-}NCO + HOH \rightarrow R\text{-}NH\text{-}CO\text{-}OH \rightarrow RNH_2 + CO_2 \uparrow \xrightarrow{RNCO} R\text{-}NH\text{-}CO\text{-}NH\text{-}R \quad (2.4)$$

This reaction is important for low density urethane foam but is avoided in RIM by careful drying of reactants and use of catalysts selective to urethane formation. Typical water level in RIM materials is $< 0.07\%$.

As Table 2.1 indicates organometallic compounds can greatly increase the rate of urethane formation. They have little effect on urea formation, but tertiary amines such as 1,4-diazabicyclo-[2,2,2]octane (DABCO) catalyze both reactions. When both catalysts are mixed there is a synergistic effect at least on urethane formation (Galla, Mascioli and Bechara, 1981). Camargo (1984) reports that the addition of 0.01mol% DABCO more than doubled the rate of a dibutyl tin dilaurate (DBTDL) catalyzed urethane RIM system. Thus most RIM formulations utilize mixed catalysts.

Isocyanates can undergo self-condensation or addition reactions. Three of the most important of these are listed in Table 2.2. Dimer formation is reversible, with dissociation being favored at elevated temperatures. Isocyanurates are stable to much higher tempera-

Table 2.1   Relative Reactivity of Phenyl Isocyanate with Various Active Hydrogen Compounds

		time in s to 25% conversion[a]
butylcarbanilate (forms allophanate)	$CH_3(CH_2)_3\text{-}O\text{-}CO\text{-}NH\text{-}C_6H_5$	3×10^5 [b]
diphenyl urea (forms biuret)	$C_6H_5\text{-}NH\text{-}CO\text{-}NH\text{-}C_6H_5$	1800[b]
water	H_2O	450[b]
2-butanol	$CH_3CH\ OH\ CH_2\ CH_3$	300[b]
1-butanol	$HO(CH_2)_3CH_3$	92
1-butanol + 0.1mol% dibutyltin dilaurate (DBTDL)		25
1-butanol + 2mol% DBTDL		6.5
1-butanol + 2mol% 1,4-diazabicyclo [2,2,2]octane(DABCO)		56
1-butanol + 0.2 DABCO + 0.1 DBTDL		~10[c]
o-toluidine	$H_2N\text{-}C_6H_4(CH_3)$	19
o-toluidine + 2mol% DABCO		7.5
aliphatic amine	$H_2N[CH(CH_3)CH_2\text{ - }O]_n\text{-}CH_2CH_3$	~10^{-3} [d]

[a] Pannone (1985; 1987) 0.5 molar NCO in dimethyl acetamide. Adiabatic method $T_0\cong23°C$, max $\Delta T\cong22°C$.
[b] Estimated from initial rate data in given by Lyman (1966, 1972).
[c] Estimated from initial rate data given by Galla et al. (1981).
[d] Pannone (1985; 1987) same as (a) but solvent was diglyme (diethylene glycol dimethyl ether).

Table 2.2   Addition and Condensation Reactions of Isocyanates

a)   **Dimerization**

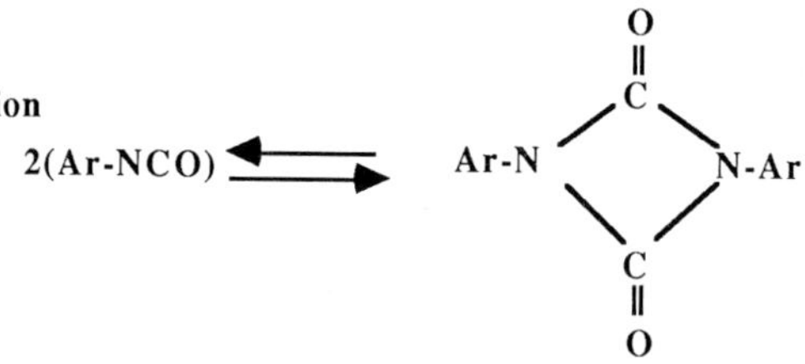

$$2(\text{Ar-NCO}) \rightleftarrows$$

Ar-N

C=O

N-Ar

C=O

b)   **Trimerization**

$$3(\text{Ar-NCO}) \xrightarrow[\Delta]{cat}$$

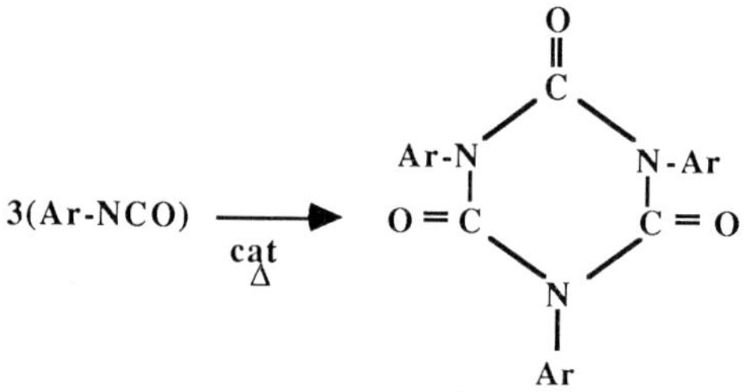

Isocyanurate

c)   **Uretonimine Formation**

$$2[\text{OCN-Ar-NCO}] \xrightarrow[\Delta]{cat} \text{OCN-Ar-N=C=N-Ar-NCO} + CO_2$$

Carbodiimide

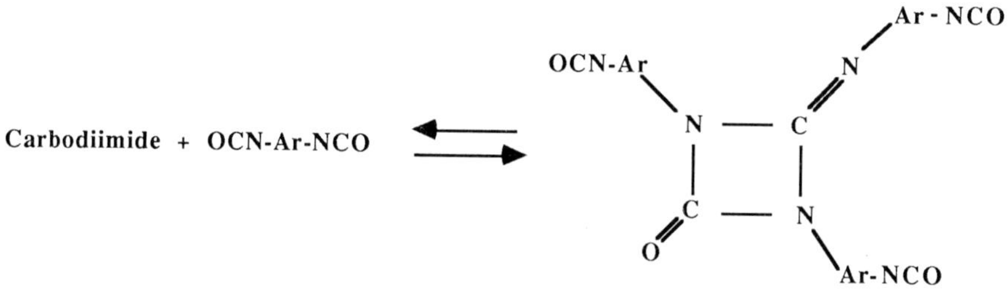

Carbodiimide + OCN-Ar-NCO $\rightleftarrows$

Uretonimine

Ar = aromatic

tures and are used in some RIM formulations (Alberino and McClellan, 1985). Isocyanurates are discussed further in Chapter 7. Thermal treatment of MDI to form some uretonimines produces a mixture that is liquid at room temperature and used extensively in RIM.

The urethane bond is also known to be reversible at high temperatures. Figure 2.2 shows how a narrow distribution urethane compound broadens in molecular weight upon heating. This rearrangement of urethane bonds can even occur in the solid state (Yang, Macosko and Wellinghof, 1986). Such temperatures may be reached during RIM polymerization and in some postcuring operations. The urea bond is believed to be more stable.

This brief review summarizes the isocyanate reactions which are important for RIM. Several much more extensive reviews of urethane chemistry are available (Saunders and Frisch, 1962; Wright and Cummings, 1969; Lyman, 1966, 1972; Reegan and Frisch, 1971). What we should notice in this overview is that urethane chemistry is well suited for RIM: urethane and urea formation give no by-products, they go to a high degree of conversion and they can be very fast. In fact the reaction of aliphatic amines with isocyanates is so rapid (note the last entry in Table 2.1) that it is very difficult to use in RIM.

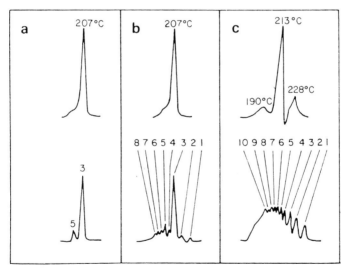

Figure 2.2   Differential scanning calorimetry (upper) and gel permeation chromatographs of the urethane compound [HO[(CH$_2$)$_4$OCONHC$_6$H$_4$CH$_2$C$_6$H$_4$HNOCO]$_3$(CH$_2$)$_4$OH (the product of 4 moles BDO with 3 moles MDI) after different thermal treatment: (a) as precipitated sample from methanol-water solution; (b) annealed at 170°C for 2 h; (c) after 50°C to 250°C DSC scan at 10°C min^{-1} followed immediately by cooling (from Yang, Macosko and Wellinghof, 1986).

2.2   Polymerization

Linear - Now that we have an idea of the important reactions used to form polyurethanes, we need to understand how the polymer structure builds up. Polyurethanes form by stepwise polymerization (also called condensation). For example a diisocyanate will couple with a diol to form a linear polyurethane

n OCN-▢-NCO + n HO ~~~~~ OH → OCN-▢ ~~~~▢ ~~~~ ▢~~~~OH   (2.5)

or, in general, OCN-[▢ ~~~~~]$_n$OH

where the junction between ▢ and ~ represents a urethane bond, -NH-CO-O-.

If the reactants are purely di-functional and there are no other reactions, then the polymer chain length, n, depends on the extent of reaction of the isocyanate groups, α, and the initial ratio of functional group concentration, r,

$$\alpha = [NCO_0\text{-}NCO]/NCO_0 \quad \text{and} \quad r = NCO_0/OH_0 \qquad (2.6)$$

With these quantities relations for the number average and weight average molecular weights can be derived from probability arguments (Macosko and Miller, 1976, using Equations 23 and 55 with $\alpha = p_B$ and r = B/A)

$$M_n = \frac{rM_1 + M_2}{1 + r - 2r\alpha} \qquad (2.7)$$

$$M_w = \frac{r(1 + r\alpha^2) M_1^2 + (1 + r\alpha^2) M_2^2 + 4r\alpha M_1 M_2}{(rM_1 + M_2)(1 - r\alpha^2)} \qquad (2.8)$$

where M_1 is the molecular weight of the diisocyanate monomer and M_2 of the diol. These equations, like those which follow, are only valid for excess hydroxyl groups (i.e. $r \leq 1$). With excess isocyanate groups we need to redefine r and α. However, excess isocyanate leads to the allophanate reaction and crosslinking (Ilavsky & Dusek, 1983).

These equations can be used to illustrate the importance of stoichiometric balance and high conversion to achieve high molecular weight in stepwise polymerizations. For example if the reaction goes to 98% completion with perfect stoichiometry, r = 1, then M_n = $50M_{no}$, where $M_{no} = M_n$ at $\alpha = 0$. But if there is 2% excess of hydroxyl groups, r =

0.98, and the reaction goes to the same conversion then $M_n = 33M_{no}$. Figure 2.3 shows similar differences for the weight average.

With RIM machines it is possible to balance the stoichiometry to within 1% and with suitable catalysts complete urethane formation can occur in seconds. But this is not enough for RIM. As indicated in the Figure 2.3, to be able to eject a molded part, solidification must also occur. One way to achieve this would be to make the polymer below its glass transition temperature. But for fast, complete reaction the polymerization must be carried out above T_g. Thus to form and then solidify the polymer in Equation 2.5 it would be necessary to first heat and then cool the mold, typically a slow process and impractical for RIM.

Crosslinking - Another appproach to structure build up is to have crosslinking occur simultaneously with polymerization. For example if a diisocyanate reacts with a triol, branched structures will form. These eventually will lead to an infinite network. In Equation 2.9 the large arrows indicate pathways along covalent bonds to infinity, i.e. to the edges of the sample.

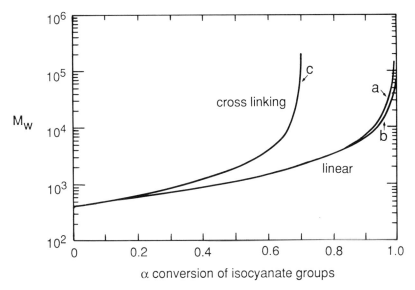

Figure 2.3   Weight average molecular weight versus isocyanate extent of reaction for a) linear, urethane formation with balanced stochiometry r =1.0 and b) 2% excess OH groups and c) a crosslinking urethane. At balanced stoichiometry the molecular weight diverges for the crosslinking polymerization at $\alpha_{gel} = 0.707$. Calculations are based on $M_1 = 250$ (4,4'diphenyl methane diisocyanate, MDI) and a 500 molecular weight diol or triol.

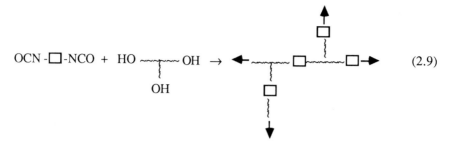

$$OCN - \square - NCO + HO \sim\!\!\!\!\uparrow\!\!\!\sim OH \rightarrow \quad (2.9)$$

$$OH$$

If all the functional groups are equally reactive then the weight average molecular weight is (Macosko and Miller, 1976)

$$M_w = \frac{r(1 + (f-1)r\alpha^2) M_1^2 + 2(1 + r\alpha^2) M_f^2/f + 4\alpha M_1 M_f}{(rM_1 + 2M_f/f) (1 - (f-1)r\alpha^2)} \qquad (2.10)$$

where M_f is the molecular weight of a polyol with a functionality f. Note that this equation reduces to Equation 2.8 for f = 2.

Figure 2.3 compares molecular weight versus conversion for linear and branched polymerizations. At about 70% reaction for the triol system, M_w becomes infinite. This means that some of the molecules are so large they are only limited by the size of the reactor or the mold. At this point the macroviscosity becomes infinite; the reaction mixture can no longer flow without tearing. Insoluble gel begins to form. Notice that the denominator of Equation 2.10 is causing M_w to diverge. From it we can derive a simple relation for the gel point

$$(r\alpha^2)_g = 1/(f-1) \qquad (2.11)$$

For simple urethane systems Equation 2.11 compares well to experimental results. For example the gel conversions of Richter and Macosko (1978) in Table 2.3 show good agreement with the theoretical value.

Equation 2.11 can be generalized to a mixture of polyols and isocyanates of different functionalities

$$(r\alpha^2)_g = (\alpha_{NCO}\alpha_{OH})_g = \frac{1}{(f_e - 1)(g_e - 1)} \qquad (2.11a)$$

where $f_e = \Sigma f_i^2 A_{fi}/\Sigma f_i A_{fi}$ is the weight average functionality of a mixture of i various functional polyols with molar concentration A_{fi}; g_e is defined similarly for the isocyanates.

Table 2.3 Gel Conversion for a Crosslinking Urethane*

Temperature (°C)	t_{gel} (min)	α_{gel}	α_{theory}
50	57.0	0.718	
65	23.5	0.690	
70	19.7	0.729	
80	12.2	0.709	
90	8.9	0.727	
		0.715(av)	0.707

*From Richter and Macosko (1978), reaction of MDI with a caprolactone triol. Gel time from viscosity rise used to calculate conversion via infrared kinetic data at the same temperatures.

Gel point relations like this are clearly important in determining mold filling limits for RIM.

If the curing temperature is above the glass transition temperature of the network, reaction will continue beyond the gel point. The reaction rate may slow down, but monomers and soluble oligomers can still diffuse through the rubbery gel to attack sites on the network. This will increase the crosslink density and thus the stiffness and strength of the network. For the same diisocyanate-triol system the soluble fraction decreases with conversion according to (Miller and Macosko, 1976)

$$w_{sol} = w_3 \left(\frac{1 - r\alpha^2}{r\alpha^2} \right)^3 + w_1 \left(\frac{1 - 2r\alpha^2 + r^2\alpha^3}{r^2\alpha^3} \right)^2 \qquad (2.12)$$

where w_1 is the weight fraction of diisocyanate and $w_3 = 1-w_1$, the fraction of triol in the initial mixture. Crosslink concentration increases following the relation

$$\mu = A_3 \left(\frac{2r\alpha^2 - 1}{r\alpha^2} \right)^3 \qquad (2.13)$$

The concentration of crosslinks is this probability multiplied by the molar concentration of triol, A_3.

The change in these quantities with conversion is illustrated in Figure 2.4. The soluble fraction decreases rapidly after the gel point. The crosslink density increases slowly just after the gel point, since there, most of the reaction is only adding pendant groups onto the gel. At high conversion each bond completes an effective crosslink. In Section 2.6 we will see how crosslink concentration can be used with rubber elasticity theory to calculate modulus evolution during curing.

The above equations for calculating molecular weight, sol fraction and crosslink density of crosslinking systems are general and can be applied to other stepwise reactions like silicones and epoxies. The statistical approach used to obtain these relations can be extended to other structural parameters (Miller ,Valles and Macosko, 1979). It can be applied to complications like unequal reactivities, e.g. toluene diisocyanate (Dusek, Ilavsky and Lunak, 1975; Miller and Macosko, 1978), substitution effects (Miller and Macosko, 1980), intramolecular cycles (Stepto 1982; Sarmoria, Valles and Miller, 1986) and oligomeric mixtures (Miller and Macosko, 1987). The statistical approach has also been extended to chainwise polymerization such as crosslinking of unsaturated polyester with styrene via free radicals (Macosko and Miller, 1976; Dusek et al., 1980; Landin and Macosko, 1984; 1988; Dotson et al., 1988).

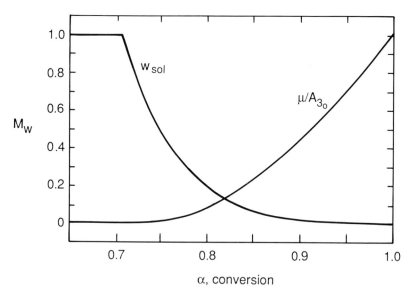

Figure 2.4   After the gel point, sol fraction decreases rapidly and the fraction of effective crosslinks, μ/A_{3_0} increases first slowly then more rapidly.

2.3 Polymerization to Form Segmented Block Copolymers

Although many urethane cast elastomers do build up their structure by crosslinking, phase separation is the primary means for building modulus in most urethane RIM formulations. One type of phase behavior was hinted at above. If the curing temperature is below the ultimate glass transition temperature of the network the reaction rate will slow down dramatically but the modulus will build up due to vitrification. Typically postcuring at a higher temperature is necessary to reach desired properties. Except for dicyclopentadiene (see Chapter 7) structure buildup by vitrification has not been utilized for RIM. Vitrification is, however, important in thermoset molding (e.g. Enns and Gillham, 1983; Williams, 1985).

The most common method of structuring in RIM is the creation of a block copolymer during polymerization. A segmented block copolymer can be made from urethanes by adding a third reactant to the system such as a low molecular weight diol, which when combined with the diisocyanate gives a copolymer which is incompatible with the other, oligomeric diol. The low molecular weight diol is often called a "chain extender" because in early urethane work it was added in a second step, after the oligomer had been reacted with the diisocyanate. The oligomer is called the soft segment because it typically has a low glass transition temperature. The reaction of diisocyanate, chain extender and soft segment diol creates a segmented block copolymer and can be represented schematically as

$$(m+1) \; OCN - \square - NCO + mHO - \bullet -OH \; + \; HO \text{\textasciitilde\textasciitilde\textasciitilde} OH \rightarrow \qquad (2.14)$$

$$OCN-\square\bullet\square\bullet\square\bullet\square\bullet\square\bullet\square \text{\textasciitilde\textasciitilde}\square\bullet\square\bullet\square\text{\textasciitilde\textasciitilde}\square\text{\textasciitilde\textasciitilde}\square\bullet\square\bullet\square\bullet\square\bullet\square\bullet\bullet- OH$$

$$or \; 1/n \; OCN - \{[\square\bullet]_m\square\text{\textasciitilde\textasciitilde\textasciitilde}\}_n \; OH$$

where m is the length of diisocyanate plus chain extender sequences in the polyurethane chain. This sequence typically has a high T_g or high melting temperature T_m and is thus called the hard segment. Hard segment sequence length is important because it controls viscosity rise during polymerization and also final polymer modulus.

Equations for the number average and weight average hard segment sequence length can be derived, again using probability arguments (Lopez-Serrano et al., 1980).

$$N_n \; = \; \frac{1}{1 - s_1 q_1^2} \quad and \quad N_w \; = \; \frac{1 + s_1 q_1^2}{1 - s_1 q_1^2} \qquad (2.15)$$

where s_1 is the molar ratio of chain extender to diisocyanate ($s_1 = m/m+1$ in Equation 2.14) and q_1, is the extent of reaction of hydroxyl groups on the extender. Equations for soft segment sequence length are the same, substituting s_2, the ratio of oligomer to diisocyanate, and q_2, the reaction of oligomer hydroxyls. Conversion of isocyanate groups is then just the sum

$$\alpha = s_1 q_1 + s_2 q_2 \tag{2.16}$$

Figure 2.5 shows how the sequence length increases as the isocyanate groups react. The middle curve shows the case of equal reactivity, $q_1 = q_2$. Note that N_w, the weight average sequence length, reaches 2 only at $\alpha = 0.65$. It would take 80% conversion for $N_n = 2$. At high conversion these values rise rapidly to $N_w = 9$ and $N_n = 5$ for $s_1 = 0.8$ (at $q_1 = q_2 = 1$).

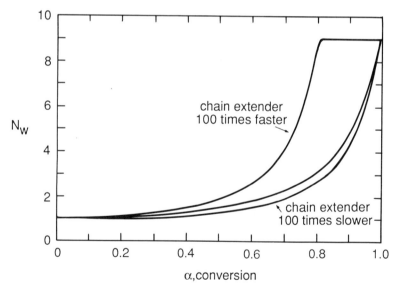

Figure 2.5   Weight average, N_w, hard segment sequence length vs. isocyanate conversion using Equation 2.15 for $s_1 = 0.80$ (i.e. m = 4, in Equation 2.14: 4 moles of chain extender, 1 mole oligomer and 5 moles diisocyanate). The upper curve is for the case where the chain extender reacts 100 times faster than the oligomer. For the middle curve they react equally and for the lower curve the chain extender is 100 times slower.

The upper curve in Figure 2.5 shows the case where the extender is much more reactive than the oligomer. Hard segment length builds earlier in the reaction. The shape of the upper curve is roughly the same as the middle one but with $\alpha \cong s_1$ stretched to $\alpha = 1$. This situation occurs, for example, when an oligomer with secondary hydroxyls is reacted with a primary chain extender.

In the opposite case, $q_2 > q_1$, which can occur, for example, when the oligomer is terminated by an aliphatic diamine, there is less difference in N_w from the equal reactivity case. However, as Figure 2.6 shows, M_w for the overall chain builds up rapidly to a plateau early in the reaction. This happens because M_w is dominated by the oligomer reactivity. In Section 2.7 we will discuss the importance of these differences for viscosity rise.

The cases shown here have equal overall stoichiometry, i.e. isocyanate equal hydroxyl concentration. Unequal stoichiometry can also be treated, but of course any imbalance will lower molecular weight as indicated in Figure 2.3. Lopez-Serrano et al. also give relations for four components and indicate how to extend their results to the case where reactivity of the second group changes after the first reacts (known as substitution effects).

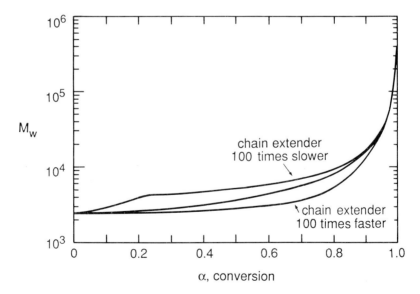

Figure 2.6 M_w, weight average molecular weight vs. conversion for the same three reactivity conditions and molecular concentrations as in Figure 2.5. The molecular weight of the diisocyanate is 250, the chain extender 90 and the oligomer 2000. Calculated from equation 55 and 58 of Lopez-Serrano, et al., (1980).

With reaction kinetic relations (Section 2.5) we can determine how hard segment length builds up with time, but viscosity and modulus evolution will also depend on nucleation and growth of the second phase. Furthermore, this hard phase can drastically alter the reaction kinetics (Section 2.7) Thus property development is more complex for polymerizations involving phase separation than with crosslinking. Before we investigate these problems let us examine some typical RIM formulations.

2.4 RIM Formulations

Since nearly all urethane RIM systems build structure by phase separation they contain at least four components: an isocyanate, oligomer, chain extender and catalysts. Let's look at each type of component more closely.

Isocyanates - Polyurethane elastomers can be formed from several diisocyanates, but the vast majority of RIM formulations are built on derivatives of 4,4' diphenylmethane diisocyanate (MDI)

$$\text{OCN}-\langle\bigcirc\rangle-\text{CH}_2-\langle\bigcirc\rangle-\text{NCO}$$

MDI is used in preference to toluene diisocyanate (TDI) because of the latter's poorer low temperature properties. Both monomers are toxic but TDI's higher vapor pressure makes it more dangerous to handle. Aliphatic isocyanates like isophorone diisocyanate have much better light stability but are more expensive and less reactive.

MDI is readily available at better than 98% purity, the impurities being the 2,4' isomer and dimers (Table 2.2). It is, however, not used in this form because its melting point is 42°C. Pure MDI also slowly forms dimer (see Table 2.2) even in its solid state. The dimer is largely insoluble in the monomer (Ferrarini, et al. 1981). To avoid these problems various methods are used to liquify MDI.

One strategy is to thermally treat MDI to produce a mixture of 85% mole MDI with 15% of its uretonimine. This forms a eutectic with a melting temperature of < 15°C making it much easier to handle in RIM equipment. Dimer formation is slower in this mixture and any dimer formed is more soluble in the uretonimine. As indicated in Table 2.2, uretonimine is in equilibrium with carbodiimide. At 50°C typical for RIM filling, there are about 5 moles of uretonimine per mole of carbodiimide. This ratio decreases to 2:1 at 100°C. Since the uretonimine has a functionality of 3 at 50°C the typical uretonimine modified MDI used in RIM has a weight average functionality $f_e = 2.15$ at 50° decreasing to 2.1 at 100°C (Equation 2.11).

Another approach to obtaining a liquid product is to react about 60% excess of MDI with a low molecular weight diol like di- or tripropylene glycol. These are called prepolymer-modified MDI's and have $f_e = 2.0$. Pure MDI can also be liquified by adding higher oligomers called polymeric MDI. Its structure is shown below where n ranges from one to ten.

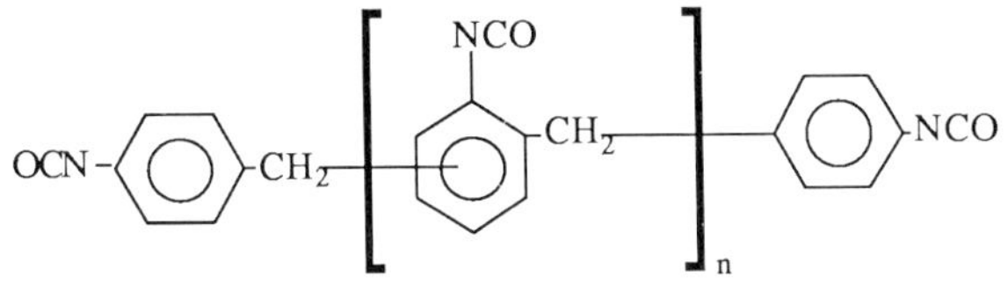

Polymeric MDI is lower cost and used mainly in rigid urethane foam formulations. Because of its branching the resulting polymer is highly crosslinked and phase separation is difficult.

While each of these approaches retains essentially the same aromatic isocyanate reactivity (Table 2.1) the different structures, uretonimine vs. diol modification, will lead to differences in the hard segments produced and thus some differences in properties (Gillis, et al., 1983). In designating formulations we will use MDI for the pure form, u-MDI for uretonimine modified and d-MDI for prepolymer or diol modified.

Oligomer - A variety of oligomers are used to form soft segments in polyurethane elastomers: polyesters, polyethers, polybutadiene (Noshay and McGrath, 1979; Speckhard and Cooper, 1986). The soft segment in nearly all current RIM formulations is based on polypropylene oxide. This oligomer is low cost and is available in large quantities with a variety of molecular weights (200 → 6,000) and functionalities ($f_e = 1-4$) because of its wide use for urethane foams. Polypropylene oxide (PPO) has a T_g of about - 60°C resulting in polyurethanes with good low temperature properties. It has a low viscosity for its molecular weight in contrast to polybutadiene or polyisobutylene oligomers. This viscosity difference is significant for RIM impingement mixing. Compared to some other available oligomers, polypropylene oxide has less water absorption than polyethylene oxide; it is not crystalline and lower cost than polytetramethylene oxide or polycaprolactones. It is also more hydrolytically stable but less oxidatively stable than polyesters.

Propylene oxide polymerizes via anionic, ring opening polymerization using alcohol as an initiator and a strong base like KOH as a catalyst (Woods, 1982).

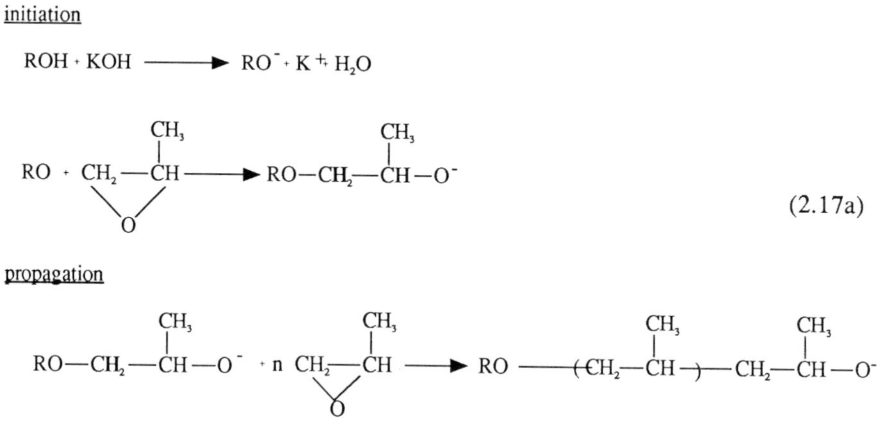

initiation

$$ROH \cdot KOH \longrightarrow RO^- \cdot K^+ + H_2O$$

$$(2.17a)$$

propagation

Chain termination is by water which generates OH⁻ or by added acid.

Abstraction of hydrogen from the monomer is an important side reaction

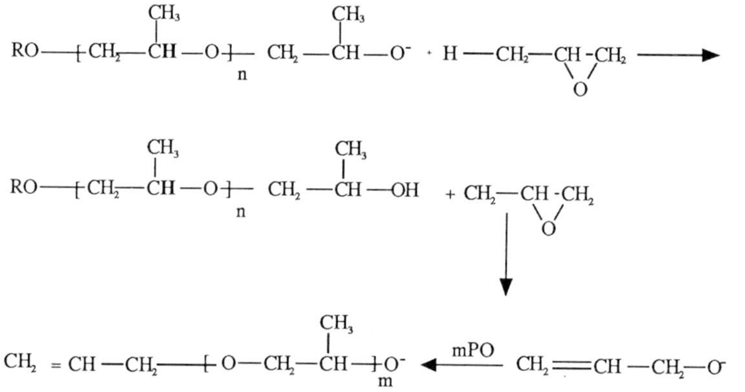

$$(2.17b)$$

The terminal vinyl group typically rearranges to allyl. "Backbiting" or abstraction of a hydrogen from a neighboring CH_3 group on the chain can also occur.

Letting the initiator ROH be a diol like propylene glycol results in α,ω, hydroxy propylene oxide. However, the chain transfer reaction (2.17b) will produce short, monofunctional species. Thus the functionalities of PPO polyols are always less than ideal and become worse at high molecular weight. $M_n = 4000$ is the typical upper limit for commercial polypropylene oxide diols and these have a functionality of 1.8 to 1.9.

Clearly by substituting a triol like glycerine or trimethanol propane for the ROH in Equation 2.17 a nominally trifunctional oligomer ($f \cong 2.6$) can be produced. Tetrafunctional and monofunctional oligomers are also available.

A major problem with polypropylene oxide is that the OH groups are secondary. As Table 2.1 indicates (compare 1-butanol to 2-butanol) a secondary OH is too slow for RIM. To create primary OH termination, ethylene oxide is substituted for propylene oxide in Equation 2.17 after a certain degree of chain buildup. It is, however, difficult to completely cap all the propylene oxide groups because ethylene oxide preferentially adds to the polyethylene oxide ends creating short PEO blocks. Low temperature reduces this tendency. Typical commercial production yields oligomers with 15-30% ethylene oxide and 70-85% primary hydroxy termination (Camargo 1984, Woods, 1982). Such ethylene oxide capped, polyproplyene oxide triols (f_e = 2.3 - 2.7) with M_n = 5-6000 are the main oligomers used today in RIM.

An alternative method for increasing reactivity is to exchange the secondary hydroxyl groups for diamines. One route is to react a polypropylene oxide diol or triol with ammonia and hydrogen over a nickle-copper oxide catalyst at 900°C and 200 atm (Yeakey, 1972). This yields termination by aliphatic amines. These react extremely fast with isocyanates (Table 2.1). Aromatic amines can be added by reacting the secondary OH groups on PPO chain ends with excess toluene diisocyanate and then converting the NCO end groups to diamines.

To increase modulus of polyurethane elastomers rigid, polymeric particles can be dispersed in the oligomer. These are made by dispersion polymerization directly in the oligomer. One approach is to copolymerize acrylonitrile and styrene in polypropylene oxide (Kuryla et al., 1966; Critchfield et al., 1972). This yields a 20% dispersion with particles between 0.25 and 0.5 μm. Another is a polyurea dispersion made by reacting a polyisocyanate with a diamine like hydrazine in polypropylene oxide (König and Dietrich, 1978; Reischl et al., 1978). Use of this in RIM formulations has been shown to improve flexural modulus by 20 to 40% (Phillips and Taylor, 1979).

Chain Extenders - It is in the chain extenders that urethane RIM systems show their greatest differences. Original formulations used 1,4 butanediol, BDO. This gave way to ethylene glycol, EG, which is cheaper and which yields polyurethanes with somewhat higher use temperature (Section 2.7). However, EG is less soluble in typical polyols, which leads to problems of phase separation in storage tanks.

Many other extenders have been tried, but the main commercial success has been with hindered aromatic diamines. As Table 2.1 indicates they have similar reactivity to catalyzed, primary OH groups. The most common diamine currently in use is an isomeric mixture of the 2,4 and 2,6 diamine isomers of 3,5 diethyl toluene (DETDA).

80%   20%

The reactivity of the para-amino group is expected to be less than the ortho and its reactivity will decrease further when the ortho group has combined with an isocyanate. Three times lower reactivity of the second group was found after the first group reacted in 2,6 DETDA (Pannone and Macosko, 1988).

DETDA has replaced ethylene glycol as the main chain extender used in RIM. Other aromatic diamines which have been evaluated for RIM include 3,3', 5,5'-tetraisopropyl 4,4'-diaminodiphenyl-methane and 1,3,5-triethyl 2,6-diamino-benzene (Nissen and Markovs, 1983; Taylor et al., 1984, Casey et al., 1985).

Catalysts and other Additives - As indicated in Table 2.1 a mixture of tertiary amine and tin catalysts is most effective for isocyanate-hydroxyl RIM formulations. All polyurea systems, i.e. diamine-terminated oligomer and diamine chain extender, require no catalyst.

Important additives include fillers, blowing agents, surfactants, internal mold release agents and pigments. Reinforcing fillers will be discussed in Chapter 8. As Chapter 6 points out, some means for expansion is needed to compensate for the shrinkage due to polymerization. This can be accomplished by dispersing dry air or nitrogen in the reactants. Silicone - propylene oxide copolymers are particularly useful surfactants for maintaining a fine dispersion of gas bubbles in the reactant tanks. For specific gravities below 0.9, physical blowing agents like Freon 11 are used.

There has been extensive development of internal mold release agents for RIM. As mentioned in Chapter 1 early RIM formulations required spraying the mold with a wax or soap after every shot. Today that is reduced to once every 20 to 100 shots by use of modified zinc stearates in polyurethanes or acidic siloxanes in polyurea formulations (Dominguez et al., 1983; Grigsby and Dominguez, 1985).

Formulations - There are three basic types of formulations which represent the major directions in RIM: urethane, mixed urethane - urea and all urea. Since formulations are proprietary it is difficult to find many specific recipes in the literature, but some recent articles and the patent literature are helpful. Table 2.4 gives some representative RIM formulations. Most have been selected because reaction kinetics (Chapter 2.5) and rheology data (Chapter 2.6 and 2.7) are available. This data can in turn be used in models for mold filling (Chapter 5) and curing (Chapter 6). Mechanical properties for several of these formulations are given in Table 6.3.

2.5 Polymerization Kinetics

In order to be useful for RIM, a polymerization must go to completion in a few minutes, preferably less than one. Thus information on reaction speed is essential for evaluating different polymerizations, for comparing catalysts and especially for relating the build up of structure (Sections 2.2 and 2.3) to time and temperature in the mold. Reaction kinetic data, combined with an understanding of the structural changes due to crosslinking or to phase separation, are necessary for modelling the rheological changes which control mold filling and curing (see Figure 2.1).

Despite numerous studies over the past forty years, the mechanisms of isocyanate reactions with active hydrogen compounds are not completely understood. An example of the complexity of these reactions is shown in the mechanism proposed by Robbins et al. (1984), for metal ion catalyzed formation of urethanes

$$\text{(2.18)}$$

Table 2.4   Some RIM Formulations[a]

Eflex Formulation	Isocyanate	Oligomer	Extender	Cat.	MPa
F-1 Crosslinked	25.3 u-MDI	74.7 PCL	- - - -	0.05 DBTDL	(Tg = 45°C)
(Manas-Zloczower and Macosko, 1987)		Mn=653 f@3.0	(Tone 305 Union Carbide)		
F-2 Prototype BDO (Camargo et.al 1985)	36.2 MDI	52 PPO-EO Mn=4080 f=1.95	11 BDO (Poly G 55-28 Olin)	0.075 DBTDL	150
F-3 RIM 2200(b) (Castro and Macosko, 1982)	35.7 u-MDI 143 g/eq (Isonate 143L,	54.6 PPO-EO Mn=5000 f@2.5, SAN filler Dow)	9.6 BDO (Niax 31-23 Union Carbide)	0.065 DBTDL	150
F-4 EG(c) (Turner et.al., 1985)	44.4 u-MDI (Isonate 143L)	46.7 PPO-EO Mn @5000 f @ 2.6 (Voranol, Dow)	8.4 EG	0.5 DBTDL	130
F-5 Urea-urethane(d) (Meyer, 1985)	30.4 d-MDI 178 g/eq	57.3 PPO-EO (Voranol, Dow) (Mondur PF, Mobay)	12 DETDA	0.15 DBTDL 0.1 DABCO	200
F-6 Prototype urea (Pannone, 1985)	31.7 u-MDI 143 g/eq	53 NH2PPO Mn=2000 (Rubinate LF168, Rubicon)	15.3 DETDA f @2.0 (Jeffamine D2000 Texaco)	- - - -	~300
F-7 Urea (Vespoli, 1985)	53 d-MDI 317 g/eq	32% NH2PPO Mn @6000 (Isonate 231)	13% DETDA f @3.0	- - - -	244
F-8 Bayflex 150	u-MDI	aromatic NH2	DETDA terminated PPO	- - - -	~300

(a) Concentrations in wt%.
(b) Similar to Mobay's Bayflex 90 and 91 formulations.
(c) Similar to Bayflex 101.
(d) Similar to Dow's Spectrim 50 and Bayflex 110.

This leads to a rate expression with an overall order which varies from 1 to 2 and is first order in catalyst. Richter and Macosko (1978) proposed a catalyst dissociation step which results in an order of 1/2 to 1 with catalyst.

Essentially all urethane reaction kinetic data in the literature fall within these limits (Camargo, 1984). This suggests a simplified kinetic expression

$$\frac{d[NCO]}{dt} = -k[C]^a[NCO]^b[H]^c \tag{2.19}$$

$$\text{with } k = A \exp(-E_a/RT) \tag{2.20}$$

where $a = 1/2 - 1$ is the order with respect to catalyst, $b + c = 1 - 2$ is the overall order of the reaction and C, NCO and H are the concentrations of the catalyst, isocyanate and active hydrogen compounds respectively.

Equation 2.19 is not a mechanistic model. It has only one rate constant with a single activation energy to express a multitude of rates and equilibrium constants as in Equation 2.18. Often the expression is further simplified by lumping the catalyst concentration into the rate constant. Frequently we consider the isocyanate concentration equal to the active hydrogen, [NCO] = [H], since most polymerizatons are run very near stoichiometry. Thus the simplest expression used to fit urethane kinetic data is

$$\frac{d[NCO]}{dt} = -k[NCO]^b \tag{2.21}$$

with b=2 most typical.

Sometimes a second rate is added for the slow reaction that occurs without catalyst (Baker and coworkers 1947, 1949, 1957; Borkent, 1974; Steinle et al. 1980). This leads to an expression with two rate constants

$$\frac{d[NCO]}{dt} = -k_1[C]^a[NCO]^b[H]^c - k_2[H]^d \tag{2.22}$$

with k_1 and k_2 in the usual Arrhenius form, (see Equation 2.20). The second rate constraint is also useful for RIM systems with two different reactions such as urethane-isocyanurate (Vespoli and Alberino, 1985a) or formulations with aliphatic and aromatic diamines (Pannone and Macosko, 1987b; Vespoli et al, 1986).

Less work has been done on urea formation, but it is believed that the product catalyzes the reaction (Baker and Bailey, 1957, Craven, 1957). This leads to an overall order from 1 to 3. Pannone and Macosko (1987a) found good third order fits to the reaction of phenyl isocyanate with aniline and o-toluidene in solution (Table 2.1).

It is a particular challenge to measure reaction kinetics during fast bulk polymerization. Titration is only useful for soluble, relatively slow reactions (e.g. Lipshitz and Macosko, 1977). Infrared, especially Fourier transform instruments, can be much faster and can monitor several chemical changes simultaneously (Camargo et al., 1985; Yang, 1987). This is particularly helpful with phase separating systems, as will be discussed in Section 2.7.

Thermal methods, based on the heat of reaction, are cruder but simpler. Differential scanning calorimetry is widely used for heat activated thermosetting systems (Prime, 1981). However, the fact that RIM systems are typically mixing activated and very fast makes DSC less useful. Hager and coworkers (1981) were able to quench a urethane RIM formulation in liquid nitrogen immediately after mixing. They obtained reasonable results upon reheating samples in a DSC.

Another thermal method, adiabatic temperature rise, uses the mixing activation and high exotherm inherent to RIM chemistry. Typically, any container over 5-10 cm diameter is sufficient to maintain adiabatic conditions in the center of a high speed RIM polymerization. Polymers are poor conductors. A 220 cm^3 polypropylene cup has worked well in our laboratory. Figure 2.7 shows our experimental setup. Thermocouples with a 0.3s response time are made by twisting or welding very fine, 0.25mm dia, copper-constantan wire. These are threaded through the walls of the cup to hold them in a central location. Everything, thermocouples, cup and tygon tubing, is disposable. The thermocouple signals are recorded at up to 10 times per second on a microcomputer. Frequently, five consecutive data points are averaged to smooth results. More details of the data collection system and a listing of the analysis program are given by Camargo (1984) and Pannone (1985).

Demounting the mixhead from the mold is time consuming with industrial scale machines. As indicated in Figure 2.7, a practical solution is to mount a thermocouple in the runner or in a thick section of the mold or even protruding from the end of the mixhead ram. Reasonable results can be obtained, especially if correction is made for heat loss. For very fast systems even the time to fill a cup is significant. Thus it is desirable to get the thermocouple as close as possible to the impingement mixing point.

Figure 2.8 shows a set of data plotted directly from a microcomputer for the prototype BDO formulation (F-2 in Table 2.4) at several catalyst levels. The adiabatic temperature rise is 130°C for the three high catalyst curves. This leads to a heat of reaction of 83.6

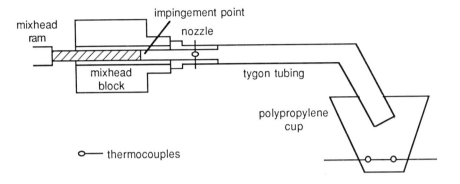

Figure 2.7   Experimental apparatus for making adiabatic temperature rise measurements. An adaptor
nozzle connects the mixhead to tygon tubing which flows the reactants into the polypropylene
cup. A thermocouple mounted in the nozzle close to the impingement point is very helpful
for measuring extremely fast reactions (adapted from Pannone, 1985).

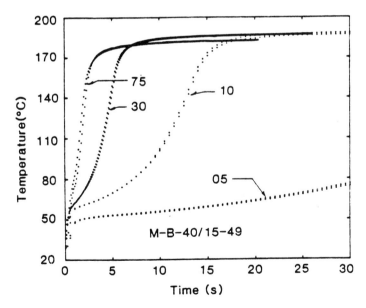

Figure 2.8   Typical adiabatic temperature vs. time for a polyurethane system based on MDI, BDO and
PPO-EO (M_n = 4000), 49% hard segment; curve labels correspond to 10^{-3}wt% DBTDL
(from Camargo, 1984).

Table 2.5   Heats of Reaction of Aromatic Isocyanate with Various Active Hydrogens

	$\Delta H_r^{(a)}$ (kJ/mol)
PI + n-butanol	83.9 ±
protype BDO, F-1 Table 2.4 [5 MDI+4 BDO+1 PPO-EO]	83.6 ± 5 [b]
PI + o-toluidine	93 ±
PI + DETDA	96 ±
PI + n-butylamine	112 ±
prototype urea, F-5 in Table 2.4 [5 MDI+4 DETDA+1 NH₂PPO]	96 ± [c] 112 ±

(a) Calculated using C_p=2.068 (Camargo, 1983).
(b) Camargo (1983); all other values from Pannone (1985).
(c) This is modeled as a two step reaction; the first value is for isocyanate with aromatic amine, the second with alliphatic.

± 5 kJ/mol (see Equation 2.25). As Table 2.5 indicates this is in good agreement with the value for n-butanol with phenyl isocyanate, PI. Other studies report similar values (Aleksandrova et al., 1970; Steinle et al., 1980; Perry, 1982; Castro and Macosko, 1982). However, Lovering and Laidler (1960) found 105 kJ/mol for the reaction of PI with a large excess of n-butanol as a solvent and Richter and Macosko (1978) give 103 for MDI with a polycaprolactone triol.

How can we obtain kinetic parameters from curves like Figure 2.8? The energy balance for the adiabatic case relates the rate of temperature change directly to heat generation due to the rate of reaction

$$\rho C_p \frac{dT}{dt} = -\Delta H_r \frac{d}{dt} [NCO] \tag{2.23}$$

where C_p is the heat capacity, ΔH_r the heat of reaction and [NCO] the isocyanate concentration. If we rearrange 2.23 and eliminate dt from both sides then an incremental change in temperature is proportional to an increment in reaction

$$dT = \frac{-\Delta H_r}{\rho C_p} d[NCO] = \frac{\Delta H_r [NCO]}{\rho C_p} d\alpha \qquad (2.24)$$

To relate ΔH_r to the maximum temperature rise we assume that 1) the reaction goes to completion $\alpha = 1$; 2) there are no other heat sources than the reaction; and 3) ρ, C_p and ΔH_r are constant. Then we can integrate 2.24 with $T = T_0$ at $\alpha = 0$ and $T = T_{max}$ at $\alpha = 1$ to give

$$T_{max} - T_0 = \Delta T_{ad} = \frac{\Delta H_r [NCO]_0}{\rho C_p} \qquad (2.25)$$

and

$$T - T_0 = \Delta T_{ad} \alpha \qquad (2.26)$$

Substituting these results back into 2.24 and assuming simple second order kinetics gives

$$\frac{T_{max} - T_0}{[NCO]_0 (T_{max} - T)^2} \frac{dT}{dt} = A \exp(-E_a/RT) \qquad (2.27)$$

This result can be used to analyze the data of Figure 2.8. A simple fitting program in the microcomputer can determine the local slopes which are shown in Figure 2.9. These plus the temperature data of Figure 2.8 allow evaluation of the left hand side of Equation 2.27. This is plotted in Figure 2.10; the straight line indicates good agreement with a second order. At low and high temperature (low and high conversion) there is some deviation. In these regions the derivative is less accurate and the simple kinetic model may also be invalid.

A simpler method for evaluating the parameters involves finding the maximum rate, $(dT/dt)_m$, and temperature at that maximum, T_{mr} (Camargo, 1984). Using T_{mr} the overall activation energy becomes for any simple, n^{th} order reaction

$$E_a = \frac{T_{mr}^2 n}{T_{max} - T_{mr}} \qquad (2.28)$$

With E_a and $(dT/dt)_m$, the frequency factor A can be found from Equation 2.27. For the 0.030 wt % catalyst sample in Figure 2.9 $(dT/dt)_m = 42°C/s$ and $T_{mr} = 135°$. From Figure

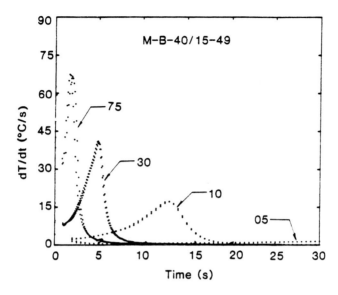

Figure 2.9   Rates of temperature rise, dT/dt, corresponding to the adiabatic rises shown in Figure 2.8 as a function of DBTDL concentration (from Camargo, 1984).

2.8 T_{max} = 184° which, for n = 2, gives E_a = 56.3 kJ/mol and A = 9.1 x 10^6 s^{-1} in good agreement with the results in Figure 2.10 and Table 2.6.

Our construction of Figure 2.10 uses a simple minimization of the differential error. More sophisticated parameter estimation methods are discussed by Camargo et al. (1983a) including integral error minimization and nonlinear regression for reaction order. However, it is more useful for comparisons to fix the order at some integer or half integer value. Corrections for a small heat loss due to conduction can be made by evaluating an overall heat transfer coefficient from the cooling of the sample after the reaction. Variable heat capacity can also readily be included if $C_p(T, \alpha)$ data is available (Steinle et al. 1980; Pannone, 1985).

Table 2.6 shows kinetic parameters for many of the RIM formulations of Table 2.4 as well as results on closely related ones. The parameters for F-2 were obtained from fitting all the data in Figure 2.8 simultaneously. In Figure 2.11 the adiabatic temperature rise predicted with these parameters is compared to the data. The fit is good except at the lowest catalyst level. Here a "two mechanism" model like that of Hager et al. (1981), which attempts to include also the uncatalyzed reaction, might give a better fit. However, for slow reactions there can be complications due to simultaneous phase separation as we shall see in Section 2.7.

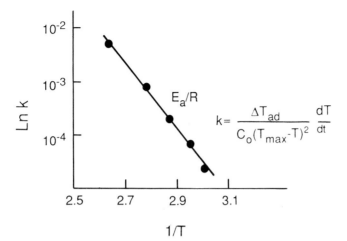

Figure 2.10   Arrhenius plot, Equation 2.27, of the data from Figure 2.9 for 0.03% catalyst assuming a second-order overall rate constant. $E_a = 55 kJ/mol$.

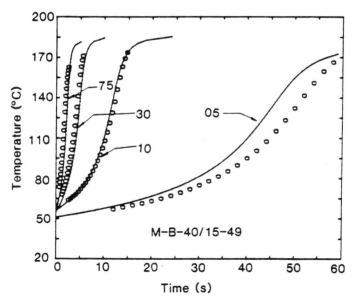

Figure 2.11   Comparison between predictions of adiabatic temperature vs time using the kinetic parameters in Table 2.6 (solid lines) and the experimental values from Figure 2.8. Note that the points at the beginning and end of the reaction were not used in the fit (from Camargo, 1984).

The last two entries in Table 2.6 are for urea formulations. Both contain amine-terminated polypropylene oxide. From Table 2.1 we expect these aliphatic amines to react several orders of magnitude faster than their aromatic partners in the same formulations.

Table 2.6   Kinetic Parameters

$$d[NCO]/dt = A_1 \exp(E_1/T)\,[C]^a\,[NCO]^b\,[H]^c + A_2 \exp(E_2/T)[H]^d \quad \text{(Equation 2.22)}$$

Formulation(1)	A_1 (2)	E_1 $-10^3\,K$	a	b	c	A_2 (2)	E_2 $-10^3\,K$	d
F-1 crosslinked (4)	3.72×10^4	4.67	-	2(3)	-	-	-	-
- several catalyst levels (5)	4.1×10^7	5.30	.75	2	-	-	-	-
F-2 prototype BDO (6)	3.24×10^9	6.59	.82	2	-	-	-	-
- PPO-EO $M_n \cong 2000$ (6)	4.35×10^8	6.69	.54	2	-	-	-	-
- PPO-EO $M_n \cong 2000$, u-MDI (7)	3.5×10^{10}	7.66	.7	2	-	5.1×10^3	4.9	2
F-3 RIM 2200 (8)	1.06×10^7	6.40	-	2	-	-	-	-
F-6 prototype urea (10)	2.0×10^4	2.06	-	2	1	6×10^3 (9)	0	3
F-7 Spectrim (11)	2.73×10^2	2.49	-	2	-	∞ (9)	-	-

Note that E_1 and E_2 are in temperature units, E_a/R. All concentrations are in mol/L.

(1)   see Table 2.4 or cited references for formulations
(2)   units to give a rate of mole/L·s
(3)   overall order only is given by most authors, thus in most cases b represents b+c.
(4)   Manas-Zloczower and Macosko (1988)
(5)   Richter and Macosko (1978, Equation 24)
(6)   Camargo (1984, Table 3.7)
(7)   Hager et al. (1981)
(8)   Castro and Macosko (1982)
(9)   for the aliphatic amine only
(10)   Pannone (1985)

Both of the studies indicated in Table 2.6 found that this is indeed the case and express this with a two step model. Pannone's second order rate constant gives a half time for the aliphatic amine of 2ms. Vespoli and coworkers (1985) found that all the aliphatic amine was converted within their shortest measurement time and therefore assumed an infinite rate. At current RIM filling speeds an infinite rate is a good approximation. This is illustrated in Figure 2.12. Here the average initial temperature of the two reactants is 56°C, but the thermocouple in the stream emerging from the mixhead (note Figure 2.7) reads 87°C during the steady state filling of the adiabatic cup. The difference, 30.6°C, agrees with the calculated value obtained from the heat of reaction for aliphatic amines with isocyanate and the initial concentration (Pannone and Macosko, 1987).

2.6   Rheology and Properties of Crosslinking Urethanes

If no phase separation occurs during polymerization, then kinetic results like those given in Table 2.6 can be combined with structural relations to predict, using Equation 2.8

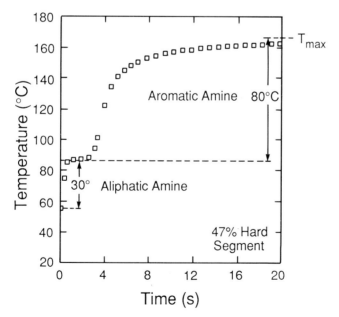

Figure 2.12   Overall adiabatic temperature rise for formulation 6 in Table 2.4 (u-MDI/DETDA/D2000 - 47% hard segment). T_0 = 55.5°C, calculated from temperatures of the two streams entering the mixhead. Flow rate Q = 100 cm^3/s. Circles: thermocouple located in tube near mixhead exit; squares: thermocouple in cup, see Figure 2.7 (adapted from Pannone, 1985).

or 2.10 for example, M_w evolution with time and temperature in the mold. With property measurements like viscosity or modulus vs. time, we can develop correlations between structural parameters and properties. Using such correlations only reaction kinetic data would be necessary to predict the viscosity and modulus build up which control the RIM cycle (see Figure 2.1). This strategy is indicated in Figure 2.13. We have already used this approach in Table 2.3 where gel conversion calculated from branching theory was compared to the gel point determined by viscosity rise and infrared measurement of reaction kinetics. Let us now see how we can extend the strategy to the entire viscosity rise curve.

For linear polymers bulk viscosity is known to depend on M_w

$$\eta \;=\; KM_w^a \tag{2.29}$$

The exponent, a, is 1-2 at low molecular weight but increases to 3.4 above a critical M_w. A theoretical explanation of this 3.4 power is still a challenge to polymer physicists (e.g. Doi and Edwards, 1986). The proportionality term, K, depends on interchain friction (Berry and Fox, 1968).

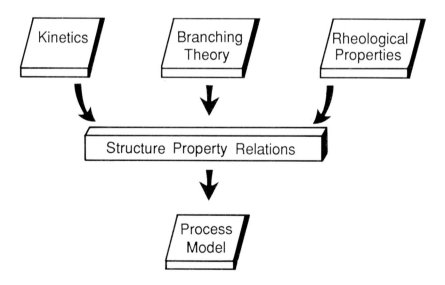

Figure 2.13   The strategy for calculating property changes during crosslinking combines reaction kinetics and property measurements with branching theory.

For branched polymers less is known. Some authors have used Equation 2.29 with values of the exponent intermediate between 1 and 3.4. It is reasonable that the M_w dependence will be less than that for a linear polymer. When a linear chain adds a monomer unit it directly increases the coil size or radius of gyration. But if a monomer unit adds to a branch the effective size of the polymer may hardly increase at all. Lipshitz and Macosko (1976) report 1.0 to 2.6 for the exponent for several crosslinking polymerizations. Figure 2.14 shows their results for a polycaprolactone triol reacting with an oligomer made from tripropylene glycol and excess 1,6-hexane diisocyanate. Viscosity was measured vs. time in a cone and plate rheometer. M_w at the same values of time was calculated from Equation 2.10 and isocyanate conversion, α, from titration measurements. We see that at 55°C viscosity depends on M_w to the 2.4 power. This exponent seems to decrease with temperature. This may be due to the fact that the T_g of this polyurethane network when cured is high, about 50°C.

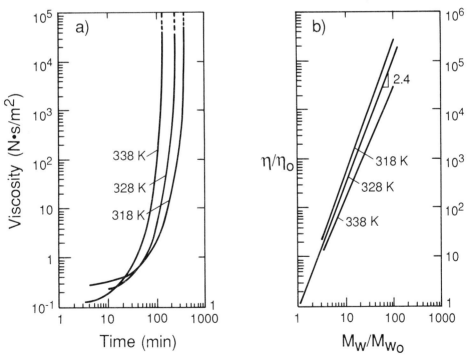

Figure 2.14   a) Viscosity vs. time for a polycaprolactone triol polymerized with an isocyanate prepolymer. b) The same data plotted against molecular weight following Equation 2.30. M_w calculated from conversion kinetics and branching theory. At 55°C the slope is 2.4 but decreases with increasing temperature (replotted from Lipschitz and Macosko, 1976).

Figure 2.14 expresses the temperature dependence of the viscosity of the branched polymer simply as the temperature dependence of the initial reactant mixture

$$\eta \; = \; \eta_0 \, (T) \, (M_w/M_{w0})^a \tag{2.30}$$

Arrhenius (Lipshitz and Macosko, 1976) and WLF type equations (Enns and Gillham, 1983; Bidstrup and Macosko, 1987) have been used for $\eta_0(T)$.

Non-Newtonian effects appear to be small for polymers built up from branched monomers or oligomers. Only very near the gel point is the molecular weight high enough for shear thinning, normal stresses and viscoelasticity to be important (Lipshitz and Macosko, 1976; Valles and Macosko, 1979a; Hickey and Macosko, 1981). As we shall see in Chapter 5 the initial viscosity rise, η_0 to $10\eta_0$, is most important for process flow problems like mold filling. With crosslinking of long chains or if particulate fillers are present, shear thinning and time dependent effects can be pronounced even at the beginning of the reaction.

A more fundamental correlation for the viscosity of branched polymers has been developed by Graessley (1982). He reviewed literature results for a number of star branched polymers and found that Equation 2.29 can still be used if M_w is replaced by gM_w where

$$g \; = \; r_{gb}/r_{gl} \tag{2.31}$$

the ratio of the radius of gyration of the branched molecule to a linear one of the same M_w.

Randomly branched polymers are produced by typical crosslinking processes. Miller and coworkers (1979) have shown how to calculate the weight average longest linear chain in a randomly branched molecule, M_{Lw}. This quantity is similar to gM_w and we have found it to give good correlations to viscosity for silicone elastomers (Valles and Macosko, 1979a; Hickey and Macosko, 1981), for epoxies (Bidstrup and Macosko, 1987) and for an acrylic system (Landin and Macosko, 1984). While this correlation has not yet been tried on crosslinking urethanes, it should also be successful.

Beyond the gel point the reacting mixture becomes a soft, weak solid. As Figure 2.1 indicates, the modulus of this solid must build high enough to permit demolding. If the reaction is carried out at a temperature above T_g, we can use rubber elasticity theory to predict how the modulus evolves with crosslinking.

The modulus of an ideal elastomer depends on the concentration of crosslinks, μ, (see Equation 2.13) and the concentration of chains between crosslinks or network chains, υ. If there are no interactions or entanglements between the network chains then

$$G = (\upsilon - \mu)RT \qquad (2.32)$$

(Graessley, 1982). The concentrations υ and μ are simply related by the connectivity of the network. For a completely reacted network with tetrafunctional junctions there will be four chains for every two junctions, $\upsilon = 2\mu$. For trifunctional junctions three chains are needed to link up every pair of junctions or $\upsilon = 3/2\mu$. Thus for a trifunctional crosslinker the modulus relation becomes

$$G = 1/2\mu RT \qquad (2.33)$$

In Equation 2.32 the $-\mu$ term represents the mobility of the crosslinks. Entanglements can suppress that mobility and can also add an additional contribution to the modulus due to interactions between the network chains. This results in

$$G = (\upsilon - h\mu)RT + G_n^o T_e \qquad (2.34)$$

where h expresses the effect of entanglements on the crosslinks, h=0 being the maximum influence, G_n^o is the plateau modulus and T_e is the probability that a pairwise interaction between two network chains is trapped (Pearson and Graessley, 1978; Gottlieb and Macosko, 1982). For trifunctional crosslinks the trapping factor becomes (Miller and Macosko, 1976; Valles and Macosko, 1979b)

$$T_e = \left(\frac{2r\alpha^2 - 1}{r^2 \alpha^3} \right)^4 \qquad (2.35)$$

We can test Equation 2.34 against networks made with polypropylene oxide triols and diisocyanates or diols with triisocyanate. This is shown in Figure 2.15. In the lower curve the networks are assumed to be perfectly connected, $T_e = 1$ and $\mu = [A_3]_o$. The non zero intercept and slope $> 1/2$ indicate the influence of entanglements. However, no network is completely connected; there are always some dangling and unattached chains.

These can be estimated from the sol fraction, Equation 2.12. Ilavsky and Dusek (1983) have done this for their networks and calculated T_e and μ. Their results are shown in the upper curve. From this we can estimate h = 0.9 and $G_n{}^\circ$ = 1.2 MPa. Gottlieb and co-workers (1981) found a similar value for h for trifunctional silicone networks. There are no plateau modulus values in the literature, but estimates from similar molecules indicate that it should be lower, about 0.5 MPa (Graessley and Edwards, 1981). Thus, although we do not yet have complete information to predict the modulus of typical polyurethane networks, it does appear that Equation 2.34 is good starting point. With α from reaction kinetics, μ from Equation 2.13, T_e from Equation 2.35 and estimates of the two entanglement parameters, h and $G_n{}^\circ$ we can predict the evolution of modulus with time and temperature during molding. Equation 2.34 can also serve as a base for modifications due to foam and to filler particles.

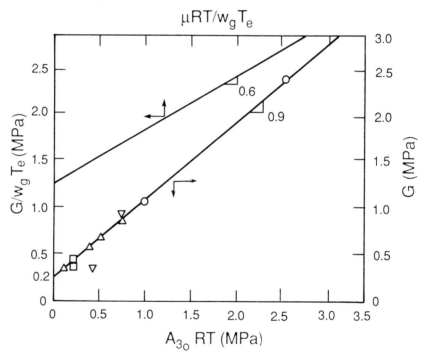

Figure 2.15   Lower curve: modulus vs. crosslinker concentration for polypropylene oxide networks cross-linked with isocyanates: ○ Ilavsky and Dusek (1983) MDI + PPO triols; △ Allen et al. (1976), MDI + mixtures of a PPO triol and diol; ▽ Smith (1985) same as Allen but with TDI; □ Sung and Mark (1981) PPO diols + triisocyanate. Upper curve: Ilavsky and Dusek data with additional points from imbalanced stoichiometry all corrected for incomplete network formation.

Rubber elasticity theory also gives a stress strain relation for elastomers

$$\sigma = G(\lambda^2 - 1/\lambda) \tag{2.36}$$

where σ is the tensile (or compressive) force over the deformed area and λ is the extension ratio, the deformed length over the initial length. Entanglement effects will add additional terms to this equation, but so far the theories have not been highly successful (Gottlieb and Gaylord, 1984). Phenomenological constitutive relations for elastomers such as the Mooney-Rivlin and Vlanis-Landel equations have been reviewed by Ogden (1986).

Gent (1978) and Smith (1978) have reviewed failure behavior of crosslinked rubber. Smith has developed failure envelopes independent of temperature and rate by plotting ultimate stress divided by ultimate extension vs. extension. He found ultimate extension proportional to $\mu^{-1/2}$. Tear energy at very low rates shows the same dependence but at high rates viscoelastic effects are important (Gent and Tobias, 1982).

Figure 2.16 shows viscoelastic properties for two of the samples of Dusek and Ilavsky (Havranek et. al, 1986). The one with balanced stoichiometry shows typical elastomer response with low loss modulus, G'', and a storage modulus, G', independent of frequency up to the glassy region. The sample with excess triol is just beyond the gel point. As we might expect from Equation 2.34, G' is much lower. Also note that $G' \cong G''$ over the middle frequency range. This is similar to the results of Chambon and Winter (1985) on silicone rubber near the gel point. These changes in viscoelastic behavior are similar to those observed for other network polymers as crosslink density is increased (Ferry, 1980; Scholtens, 1984).

As we have have mentioned several times urethanes are typically crosslinked above their glass transition temperature. However, as Figure 2.16 indicates (shift in G'' peak) we would expect the glass transition to increase with conversion. It is well known for linear chains that T_g increases with chain length; the chain ends can be thought of as plasticizing units.

$$T_g = T_{g\infty} - K/M_n \tag{2.37}$$

where $T_{g\infty}$ is the glass transition for an infinitely long chain and K is a constant, about 10^5 K for polystyrene (Flory, 1953; Kovacs, 1963). Branch points and thus crosslinks decrease chain flexibility nearby and therefore raise the glass transition (Rietsch et al., 1976).

Networks formed by condensation reactions like polyurethanes incorporate a crosslinking molecule which will have a copolymer effect on T_g. Some relations have been developed for T_g changes during crosslinking, but as yet none have successfully combined all of these effects. As an example of the changes we might expect, T_g should increase from about 210 (pure PPO) to 330K moving along the lower curve in Figure 2.15 based on measurements on similar samples by Andrady and Sefcik (1983). Feger and MacKnight (1985) report T_g rising from 207 to 235 during the polymerization of a PPO diol with a triisocyanate.

2.7 Rheology and Properties of Phase Separating Urethanes

In crosslinking polyurethanes the polymer molecular structure builds directly from bond formation and this in turn governs the property changes like viscosity and modulus. If the reaction occurs well above the glass transition, there is only one kinetic process--

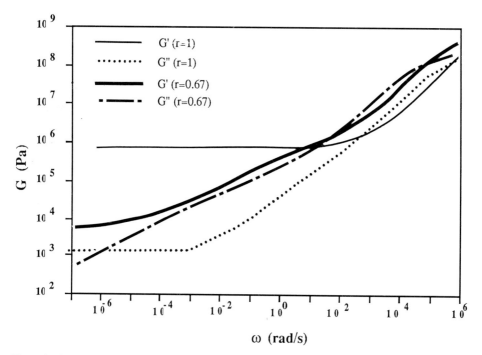

Figure 2.16   Master curves of dynamic moduli vs. reduced frequency for two samples from Ilavsky and Dusek (Havranek, et al., 1986).

covalent bond formation. Through it all property changes can be related directly to conversion. This is discussed further in Chapter 6 (see Figure 6.18).

For the polymers which predominate in RIM an additional kinetic process is important--phase separation. Not only does phase separation control property development in RIM polyurethanes, ureas and nylons, but it is highly coupled to the reaction kinetics. If the polymerization is slower than the phase separation rate, hard blocks can come out of solution, trapping monomers and oligomers. This will lead to a low molecular weight, brittle final polymer. If the polymerization is much faster than phase separation, then high molecular weight polymer will form before any phases have time to precipitate. This can slow down the phase separation rate considerably.

Clearly polymerization combined with phase separation is a complex process. To begin to understand it let us first consider what causes phase separation: the thermodynamics of segmented block copolymers. We will then look at the kinetics of this process for pre-made polyurethanes, then phase separation during polymerization. Finally, we will consider how phase separation and polymerization combine to control viscosity, modulus and strength of the growing and structuring polymer. We will not be able to come up with as complete a theoretical and experimental understanding as for crosslinking polymerizations, but hopefully we can capture the important qualitative ideas.

Thermodynamics - In Section 2.3 we saw that a diisocyanate combines with a chain extender, such as a short diol or diamine, to form the hard block in the segmented copolymer. These separate into domains as shown schematically in Figure 2.17. The driving force for this separation is the difference in solubility between the hard and soft blocks. Table 2.7 gives solubility parameters for some urethane materials. These have been calculated by different authors using different methods so there may be some inconsistencies, but the general trend is clear: typical polyurethane hard segments all have higher solubility parameters than the soft segments.

The onset of phase separation can be predicted using the Flory-Huggins relation (Flory, 1953). For a mixture of two polymers A and B

$$\chi_c = \frac{1}{2} \left(\frac{1}{\sqrt{N_A}} + \frac{1}{\sqrt{N_B}} \right)^2 \qquad (2.38)$$

N_A is the number of repeat units on the A chains. χ is a measure of the interaction between segments of A and B and can be estimated from their solubility parameters.

$$\chi \cong \frac{V_T}{RT} \left(\delta_A - \delta_B \right)^2 \tag{2.39}$$

where V_r is a reference volume based on the molecular weight of an A repeat unit and the density of polymer A; χ_c is the critical value for the onset of phase separation.

Let's consider some examples. If B is a solvent, $N_B = 1$ and $N_A \gg 1$ then Equation 2.38 gives $\chi_c \cong 1/2$, the classic result of Flory (1953). If both A and B are polymers and have the same chain length then (Scott, 1949)

$$\chi_c (N_A + N_B) = 4 \tag{2.40}$$

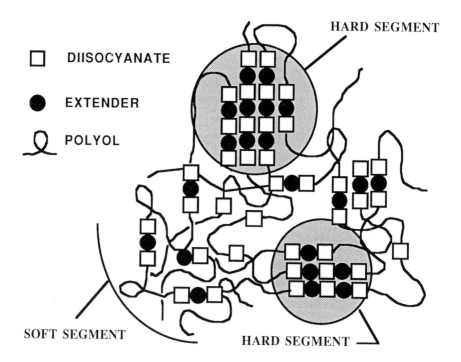

Figure 2.17   Schematic representation of phase separation in polyurethane segmented block copolymers.

Since N_A is typically > 100, χ_c is very small, < 0.1. This says that for two polymers to be mutually compatible their solubilities must be very close. And in fact there are very few miscible polymer- polymer pairs.

If one end of each chain is tied to the other then we have an AB block copolymer. This makes the pair more compatible. Leibler (1980) has treated this problem for narrow distribution of block lengths. Figure 2.18 shows χ_c $(N_A + N_B)$ for various block lengths ratios. For $N_A = N_B$, f = 0.5 from the figure we see that

$$\chi_c \, (N_A + N_B) \; = 10.5 \qquad\qquad\qquad (2.41)$$

Table 2.7   Solubility Parameters for Urethane Materials at 25°C

Hard Segments	δ $(cal/cm^3)^{1/2}$	Reference
MDI	9.8, 10.6	a, c
(MDI-BDO)$_n$	13.2, 11.5	b, c
(MDI-EG)$_n$	10.3, 11.6	a, c
(MDI-DETDA)$_n$	12.4	c
Soft Segments		
poly propylene oxide (PPO)	8.6, 9.1	a,b
poly ethylene oxide (PEO)	8.9	a
poly butylene oxide (tetra methylene oxide) (PTMO)	8.7	b
poly caprolactone (PCL)	9.1	c
1,4 polybutadiene (PBD)	8.3	b
poly dimethyl siloxane (PDMS)	7.5	b

(a) Nishimura et al. (1986)
(b) Camberlin and Pascault (1984)
(c) Ryan (1988); Ryan et al. (1988)

Benoit and Hadziioannou (1988) have extended Leibler's derivation to $(AB)_n$ segmented block copolymers. As we might expect with both ends attached χ_c increases but not great-ly. For $N_A = N_B$ and $f = 0.5$

$$\chi_c (N_A + N_B) \cong 15 \qquad (2.42)$$

The shape of $\chi_c (N_A + N_B)$ versus f should be similar to that for AB block copolymers in Figure 2.18.

We can apply these ideas to determine whether a given hard segment-soft segment pair will phase separate. For example, first consider the blend of a polypropylene oxide soft segment with MDI. From the solubility parameters in Table 2.7 $\delta_A = 9.8$ and $\delta_B = 8.6$ $(cal/cm^3)^{1/2}$ and assuming $V_r = 250$ cm^3/mol, Equation 2.39 gives $\chi = 0.61$ at 298K.

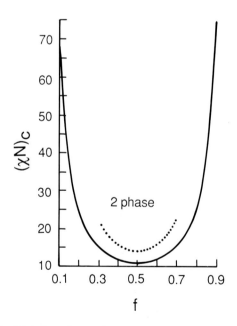

Figure 2.18   Ideal phase diagram for an AB block copolymer (Liebler, 1980) and for an $(AB)_n$ segmented block copolymer where $n \to \infty$ (Benoit and Hadziioannou, 1988).

For 2000 molecular weight PPO N_B = 34.5 while MDI is equivalent to about 4 PO units or N_A = 4. Using these values Equation 2.38 gives χ_c = 0.34. Thus we would expect this blend to be incompatible and in fact MDI is not soluble in PPO.

If we make a segmented block copolymer from the same pair, the fraction of hard segment is small, f = 0.1. Using Figure 2.18 gives χ_c ($N_A + N_B$) ≅ 70 for an AB block copolymer and about 100 for a segmented one. This leads to χ_c = 2.6. Thus, this segmented block copolymer should not be phase separated at 298K. This also confirmed by experiment (Christenson et al., 1986).

However, if we make a copolymer with a larger hard segment, for example MDI-BDO-MDI, then we do get a two phase polymer. In this case N_A ≅ 9. Thus f = 0.2 and χ_c ($N_A + N_B$) ≅ 43 or χ_c ≅ 1.0 for a segmented block copolymer. If we use δ = 10.3 for this hard segment, then from Equation 2.39 χ = 1.6. Thus phase separation is expected to occur.

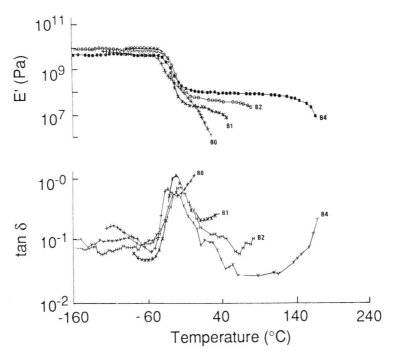

Figure 2.19   Dynamic shear modulus and loss tangent vs. temperature for a series of polyurethanes with increasing hard segment length and relatively narrow segment length distribution (adapted from Christenson et al., 1986). B0 has no BDO unit between the soft segments only one MDI; B1 has one BDO unit; B2 has two and B4 has four.

The difference between these segmented block copolymers can be seen in their dynamic modulus vs temperature data, Figure 2.19. The MDI + PPO sample (B0) shows no plateau while the MDI-BDO-MDI + PPO sample (B1) does. Note that as the hard segment length is increased the softening temperature which should relate to phase mixing also increases. This is at least in qualitative agreement with block copolymer theory. Leung and Koberstein (1986) have studied a series of similar polyurethanes by DSC annealing and found that the phase mixing temperature also increases with hard segment length.

The theory for phase separation in block copolymers has been derived for monodisperse segment length. Typical condensation methods yield a broad distribution; $N_w/N_n \cong 2$ (see Equation 2.15). Broad distribution samples show lower phase mixing temperatures and poorer phase segregation (Harrell, 1969; Ng et al., 1973; Eisenbach et al., 1985). This is illustrated in Figure 2.20. The lower phase mixing temperature can be understood by considering a broad hard segment size distribution sample as a mixture of many f values around the mean; it represents a region along the critical line in Figure 2.18. The lower f value portions of the broad distribution sample will have a higher χ and thus lower phase mixing temperature.

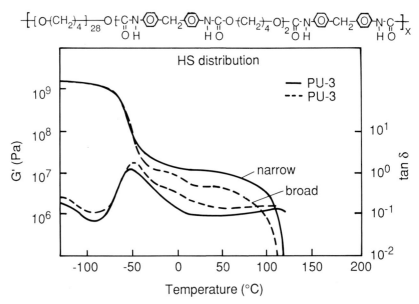

Figure 2.20   Dynamic shear modulus vs temperature comparing a narrow to a broad hard segment distribution polyurethane with the same average segment length (replotted from Eisenbach, Baumgartner and Gunther, 1985). Polytetramethylene oxide soft segment $M_n = 2000$ and MDI-BDO hard segment with two BDO units.

There are other results in the literature on segmented polyurethanes which are in qualitative agreement with block copolymer theory. Seefried, Koleske and Critchfield (1975) found that as they decreased the molecular weight of a polycaprolactone soft segment from 2000 to 830 (N_B = 18 to 7) with f = 0.5 the polymer became single phase. Zdrahala and Critchfield (1981) found softening temperature to increase as PPO soft segment molecular weight was increased from 1000 to 4000. Camberlin and Pascault (1984) made polyurethanes at f ≅ 0.3 with different types of soft segments. As they decreased δ_B (see Table 2.7) the degree of phase separation increased linearly with χ, see Figure 2.21. Phase separation also improved with soft segment length.

Kinetics - The thermodynamic arguments given above tell us whether a block co-polymer will phase separate under *equilibrium* conditions. Because these polymers are highly viscous, the kinetics of phase separation is typically also important in the formation of the final structures.

When cooled from a single phase into the two phase region, polyurethanes take 10 minutes to many hours to reach equilibrium (Wilkes and Widnauer, 1975; Kwei, 1982; Camberlin and Pascault, 1983, 1984; Koberstein and Russell, 1986). RIM polyurethanes and ureas are typically formed at temperatures where at equilibrium the polymers would be

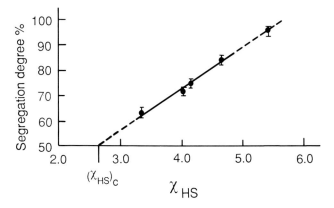

Figure 2.21   Degree of phase separation calculated from the change in heat capacity on passing through T_g of the soft segment. χ calculated from values in Table 2.7. For all samples f ≅ 0.3 (replotted from Camberlin and Pascault, 1984).

two phase, i.e. where $\chi > \chi_c$. Thus phase separation and polymerization can occur simultaneously and very rapidly. Typical techniques used to measure phase behavior like small angle X ray scattering, SAXS, are not fast enough to follow these changes (Wilkes and Emerson, 1976). However a less direct method, Fourier transform infrared, FT-IR can be used.

When urethane hard segments associate with each other hydrogen bonds form between the carbonyl oxygen and the amine hydrogen (Camargo et al., 1983b; Coleman et al., 1986). This results in a frequency shift in the C=O peak as illustrated in Figure 2.22. During polymerization the C=O peak increases as isocyanate groups convert to urethane. When the area under the free C=O peak and the hydrogen bonded one are plotted vs time, the free C=O typically goes through a maximum as shown in Figure 2.23.

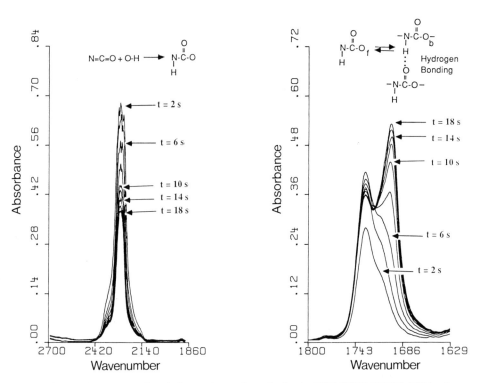

Figure 2.22   FT-IR spectra vs time during urethane polymerization (at 75°C, 0.02% DBTDL48% hard segment similar to F-2, Table 2.4). a) isocyanate disappearance b) carbonyl formation followed by frequency shift due to hydrogen bonding (from Yang and Macosko, 1989).

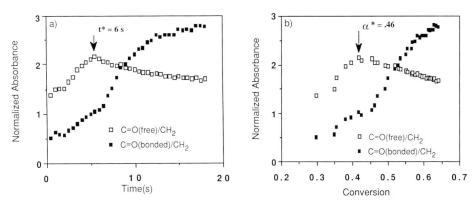

Figure 2.23 Peak areas from data similar to Figure 2.22 plotted vs. a) time and vs. b) conversion. C=O.F is the normalized area under the free carbonyl peak 1730 cm^{-1} and C=O.B is the bonded carbonyl 1705 cm^{-1} (from Yang and Macosko, 1989).

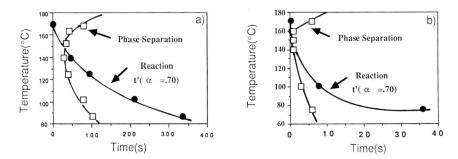

Figure 2.24 Time to reach 70% conversion and time for onset of phase separation (peak in Figure 2.23) vs temperature for a) uncatalyzed and b) catalyzed with 0.02 wt% DBTDL. Data from FT-IR similar to Figure 2.22, for a 48% hard segment MDI-BDO-PPO polyurethane (from Yang and Macosko, 1989).

The maximum in free C=O is a convenient indicator of the onset of phase separation. In Figure 2.24 the time to this maximum and time to 70% conversion are plotted for various temperatures. As temperature increases the reaction kinetics are always faster, however phase separation kinetics (the time to the free C=O maximum) slow down at high temperature. This is logical since at high enough temperature we expect phase mixing. At 50% hard segment Leung and Koberstein (1986) report an equilibrium phase mixing temperature of 185°C. In Figure 2.19 for sample B4 (44% hard segment) the modulus drops at about 170°C. These results agree with the trend in Figure 2.24; above approximately 180°C it appears that there will not be phase separation.

At lower temperatures phase separation occurs before the reaction reaches 70% conversion, before the polyurethane becomes a high polymer (see Figure 2.6). Tirrell, Lee and Macosko (1979); Camargo and coworkers (1982, 1985); Speckhard and Cooper (1986) have all shown that this premature phase separation leads to low molecular weight, brittle materials.

The addition of catalyst, Figure 2.24b, allows polymerization to high molecular weight at lower temperature. Catalyst accelerates the reaction kinetics. Phase separation speed also increases but not as much. These ideas are important for optimizing RIM mold temperatures as discussed further in Chapter 6.

Morphology - When polyurethane segmented block copolymers phase separate what shape do the phases take? There has been considerable controversy in the literature, but it appears that the basic morphology of the hard segment domains is lamellar (Briber, 1983; Briber and Thomas, 1983). The reason for the controversy is that domains are very small and thus difficult to study by electron microscopy or small angle X-ray scattering (SAXS). The length of an MDI-BDO repeat unit in its typical crystalline form is 1.7 nm (Briber and Thomas, 1985). For a 50% hard segment polyurethane the number average sequence length is about 5 which means a segment length of 9 nm. This sets at least one dimension of the lamellar domains.

Figure 2.25 represents some of the electron micrographs on both microtomed and solution cast polyurethanes with higher hard segment content (50-80%; Chang et al. 1982; Briber and Thomas, 1983; Kolodziej et al., 1986; Nishimura et al., 1986). Lamellar hard segment domains about 10x50 nm can form into fibrous, branched structures. These regions are birefringent spherulites. There are, however, always large regions where no structures can be observed indicating domains less than 10 nm. This featureless morphology seems to give ductile behavior while the highly birefringent, spherulitic morphology is more common with brittle, low molecular weight polymers. Hard segment rich, amorphous globules in the 5 μm size scale have also been observed. There is some indication that these are due to incomplete mixing (Kolodziej et al., 1986).

Small angle x ray (SAXS) studies indicate that average hard segment domain size increases with the number of repeat units up to about 6 or 10 nm but then is about constant for longer hard segments. This indicates that the hard segments are folded within the domains (Koberstein and Russell, 1986).

The aromatic polyureas used increasingly in RIM show good phase separation by dynamic mechanical tests, however no lamellar, crystallinity or spherulites have yet been observed (Camargo, 1983; Dominguez 1984; Willkomm et al., 1988).

Viscosity - The domains that form during phase separation will lead to a rapid buildup of viscosity and gelation much like a crosslinking polyurethane, although these polymers are linear. Figure 2.26a shows viscosity vs time data for a 47% MDI-BDO-PPO polyurethane. This is an uncatalyzed formulation so the reaction and viscosity changes are relatively slow. The curves are similar to Figure 2.14a. As in Figure 2.14a at higher temperature the initial viscosity is lower, but it rises faster due to the faster reaction.

When the data are plotted as reduced viscosity vs. isocyanate conversion, all temperatures reduce roughly to the same curve as shown in Figure 2.26b. Castro, Perry and Macosko (1984) found this type reduction could be made at different hard segment levels in the

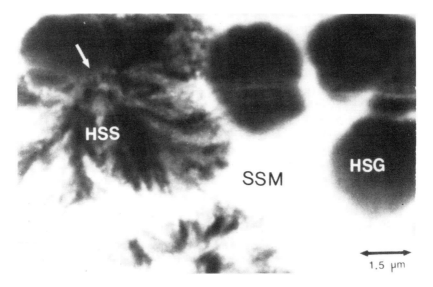

Figure 2.25 Transmission electron micrograph of 50% hard segment MDI-BDO-P(PO-EO) polyurethane microtomed. Lamellar domains form into fibrous, birefringent spherulites (HSS). The featureless regions (SSM) are also phase separated but on a finer scale < 10 nm. There are also hard segment globules (HSG) (Chang et al., 1982, photo from Chang, 1984).

58   Polyurethanes

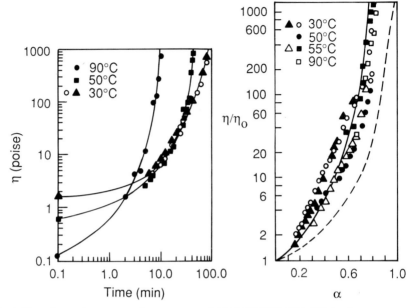

Figure 2.26   a) Viscosity vs. time at several temperatures. MDI-BDO P(PO-EO) 47% hard segment, uncatalyzed. b) Same data plotted as reduced viscosity vs. conversion. Solid curve Equation
2.43, dashed curve Equation 2.30 with $M_w^{3.4}$(Castro, Perry and Macosko, 1984).

same temperature range 30-90°C and with no or low catalyst levels. Viscosity rises much
faster than molecular weight even to the 3.4 power, Equation 2.30. This indicates that
chain entanglements and branching are not the main cause of viscosity increase.

An empirical equation which does fit the data is

$$\eta = \eta_o(T) \left(\frac{\alpha_g}{\alpha_g - \alpha} \right)^{C_1 + C_2\alpha} \tag{2.43}$$

where $\eta_o(T) = A_\eta \exp(E_\eta/RT)$. Values of the constants are given in Table 2.8; α_g is the
conversion at physical gelation due to formation of a network of hard segment domains. It
is the critical parameter for fitting the rising viscosity. Equation 2.43 indicates that α_g is independent of temperature. This can only be true over a limited range. At high temperatures
we expect α_g to increase and eventually even disappear as indicated by the phase separation
time in Figure 2.24.

Equation 2.43 says that viscosity depends on the reaction kinetics through α, but in fact it must depend on phase separation kinetics. Its success can be attributed to the narrow ranges of temperature and catalyst level over which it has been tested. Over these limited regions the two kinetic processes parallel each other, see Figure 2.24.

One of the problems with obtaining rheological data at higher temperatures and catalyst levels is the speed and heat of the reaction. To mix and load fast reacting samples essentially requires using a small RIM machine to fill the rheometer. Since most RIM polymerizations are highly exothermic (Table 1.2), the rheometer gap must be very thin to maintain an isothermal sample (Perry, Castro and Macosko, 1985). Due to start up and shut down transients mixing and precise delivery of very small samples is a challenge for current RIM equipment (see Chapter 3).

An alternative is to use a large gap, insulate the rheometer and measure under adiabatic conditions. This has been done by Blake and coworkers (1987) and some of their results are shown in Figure 2.27. Note the effect of catalyst. Viscosity rises earlier in time with the higher catalyst level, but when the data are plotted vs conversion, removing the effect of catalyst on reaction kinetics, the higher catalyst has a higher α_g. This is as we would anticipate from Figure 2.24; high catalyst is able to accelerate reaction kinetics more than phase separation.

Even the adiabatic viscometer has limitations. It is not possible to study viscosity rise at low temperatures similar to the conditions very near mold walls. The adiabatic condition means that reactions will be considerably faster than they are isothermally at T_0. It also takes some time to fill and equilibrate the large gap device. Blake's couette design is limited to reaction times > 2s. Foaming also can make the bob height uncertain.

Table 2.8   Parameters Used in Viscosity Rise Expression, Equation 2.43[a]

Hard Segment Content	34%	47	55.5	64
E_η(J/mol)	2237	2232	2220	2214
A_η MPa·s	3.095	2.764	2.608	2.405
α_g	0.93	0.88	0.85	0.82
C_1	3.5	3.5	3.5	3.5
C_2	-2.0	-2.0	-2.0	-2.0

[a] For MDI-BDO-P(PO-EO) polyurethane from Castro, Perry and Macosko (1984).

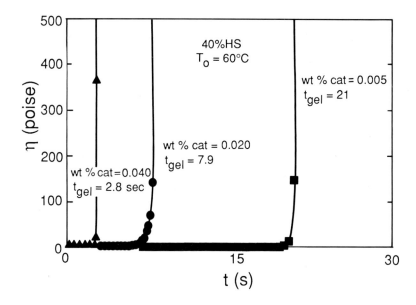

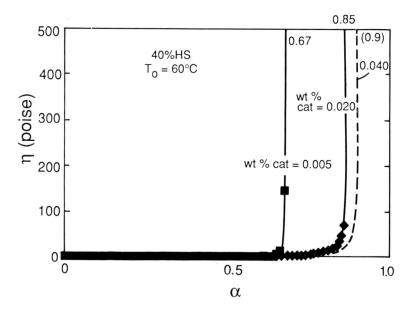

Figure 2.27   Adiabatic viscosity rise at three catalyst levels a) vs time b) vs conversion (calculated from the adiabatic temperature rise). For an MDI-BDO-P(PO-EO) 40% hard segment formulation (adapted from Blake et al. 1987).

Another approach to characterizing viscosity changes during fast RIM polymerization is to measure pressure rise during mold filling (Lawler and Gill, 1981; Castro and Macosko, 1982; Vespoli and Alberino, 1985b; Vespoli et al. 1986). If the mold cavity is simple, a long round runner or a wide thin rectangular plaque, then pressure is directly related to viscosity at each position in the mold (Equation 5.1). This problem is discussed further in Chapter 5 and some pressure rise data are shown (Figures 5.6, 7, 9, 10, 13). It is very difficult to back out an unknown viscosity function from the coupled effects of non-homogeneous temperature and conversion fields. However, two parameters can be accurately estimated: $\eta(T_0)$ and α_g. The initial slope of pressure vs time gives a good measure of the initial viscosity, $\eta(T_0)$. This could include the effect of a very fast reaction like aliphatic amine with isocyanate which is essentially complete as the mixture enters the mold (Vespoli et al., 1986). The gel conversion, α_g, can be determined from the point where pressure rises strongly, using the assumption of adiabatic filling and residence time distribution discussed in Chapter 5.

Mechanical Properties - Beyond the gel point the polymer is a soft weak solid. As phase separation continues and domains interconnect, modulus and strength increase. Figure 2.28a shows modulus development during polymerization at 100°C for several hard segment level polyurethanes. Notice that increasing hard segment level decreases α_g as also found by viscosity rise. Modulus keeps rising at long times compared to the reaction. In this region the reaction conversion has little influence on modulus; if anything, it builds high molecular weight which will *slow* down phase separation kinetics. However, the rate of modulus buildup is faster than if the sample were cooled rapidly from the melt to this temperature. This indicates, as suggested by Figure 2.24, that phase separation and phase growth begins before the polymer is completely formed. Figure 2.28b shows the effect of catalyst and temperature on modulus buildup for the same MDI-BDO system. Note that modulus builds more slowly and to a lower final level at high temperatures.

As yet, like for viscosity, there is no model which connects the buildup of modulus with phase separation and domain structuring during or even after complete polymerization. This buildup controls the time to demold a RIM part, typically 30% or more of the molding cycle (note "cure" in Figure 3.3). The role of mechanical properties in demolding is discussed further in Chapter 6. To construct rheological models further development of block copolymer theory and more data are needed.

Although there is little understood about how mechanical properties build up with polymerization and time, there is extensive data on final properties of segmented polyure-

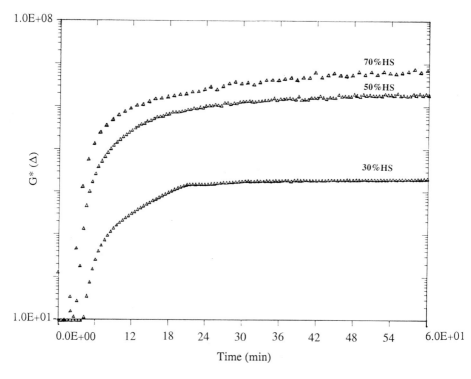

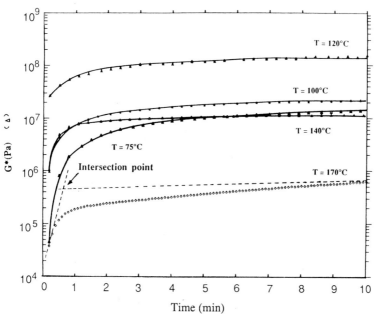

thanes as a function of various formulation parameters. The formulation parameters which can affect phase separation of segmented polyurethanes include chemical nature of the hard and soft segment, individual segment length and segment length distribution, intra-and interdomain interaction such as hydrogen bonding, hard segment content, overall molecular weight and molecular weight distribution, as well as the nature of the domain interface and the mixing of hard segments in the soft phase. The thermal treatment and mechanical history are also of importance to the final properties of the polymers. Some of the factors will be briefly discussed below.

With hard segment content being the same, longer soft segment sequence length improves the degree of phase separation, hence increases the flexural modulus as shown in Figure 2.29. The reinforcing effect of the hard segment is also clearly demonstrated as the flexural modulus increases with hard segment content at a given soft segment molecular weight. Lower soft segment T_g and flatter rubbery plateau modulus with increasing soft segment molecular weight are other indications reported in literature for improved phase separation. The effect of different hard segment length and distribution have been discussed in the thermodynamics section (Figure 2.19 and 20). Polyurethanes based on polyether polyol exhibit better properties than the counterparts based on polyester polyol. This is attributed to less hydrogen bonding between the hard segment and the ether groups in polyether soft segment in contrast to the ester groups.

Better phase separation can be achieved when the intra-domain cohesion of hard segments is increased. High modulus and low heat sag properties are often promoted by hard segment crystallinity and rigid bulky groups. Crystallization provides additional driving force for phase separation besides the thermodynamic incompatibility. In glycol extended polyurethanes, crystallinity is identified as an important mechanism for property development. Polyurethanes with amorphous hard segments either made of 2,4'-MDI and BDO or 4,4'-MDI and MDEA, show extensive phase mixing and poor mechanical properties (Camargo, et al., 1985). Figure 2.30 shows the dynamic modulus vs temperature data for typical glycol extended systems. The ethylene glycol extended system exhibits higher softening temperature and flatter rubbery plateau modulus than the butanediol extended system. This is probably due to its higher hard segment transition temperature.

Figure 2.28a Complex modulus vs time for three hard segment levels, MDI-BDO-P(PO-EO) uncatalyzed, 100°C, 0.5% strain, 6.28rad/s (from Yang, 1987).

Figure 2.28b Complex modulus G* vs reaction time at several reaction temperatures for same system as Figure 2.28a, but containing 0.02% DBTDL catalyst. 1% strain 6.28 rad/s (from Yang, 1987).

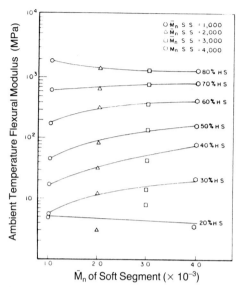

Figure 2.29   Flexural modulus at 25°C vs soft segment molecular weight and hard segment content for MDI-BDO-P(PO-EO) polyurethanes (from Zdrahala and Critchfield, 1981; also Zdrahala et al., 1979)).

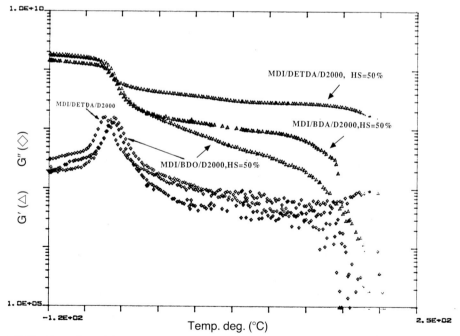

Figure 2.30   Dynamic modulus vs temperature for three hard segment types all at 50% hard segment concentration (from Yang, 1987).

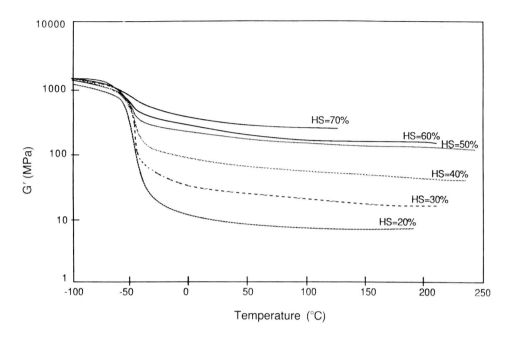

Figure 2.31   Dynamic modulus vs temperature at various hard segment content for MDI-DETDA-D2000 polyureas (from Chen, Yang and Macosko, 1987).

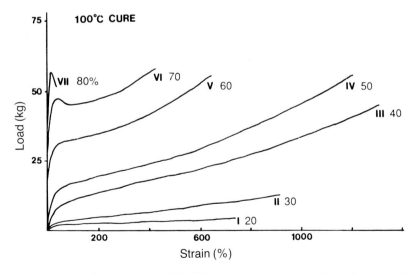

Figure 2.32   Load vs. strain for MDI-BDO-P(PO-EO) polyurethanes at various % hard segment (from Chang et al., 1982).

When diamine chain extender is used, however, even without the benefit of crystallinity, the properties are much improved over the glycol extended polyurethanes as demonstrated in Figure 2.30. Higher polarity difference between hard and soft segments and three dimensional hydrogen bonding, made possible by the extra NH group in the urea linkage, have been proposed to account for the property improvement (Bonart et al., 1974; Born and Hespe, 1985). The rigid, bulky aromatic structure of DETDA provides additional contribution to the higher modulus and higher softening temperature. Due to its much better properties, DETDA has replaced butanediol and ethylene glycol to become the dominant chain extender used in RIM. Further improvement on the properties can be achieved when amine-terminated polyol (PPO-NH2) is introduced to the formulation. These polyureas have been successfully produced via RIM and are reported to be much superior to their polyurethane and polyurethane-urea counterparts (Dominguez, 1984). Polyureas have higher thermal stability and modulus than conventional polyurethanes and they have become strong candidates for external body panel formulation in automotive industry. Figure 2.31 shows the dynamic modulus vs temperature data of polyureas based on MDI/DETDA/PPO-NH_2. The reinforcing effect of hard segment content on the rubbery plateau modulus is very similar to that of conventional polyurethanes as demonstrated in Figure 2.29.

Figure 2.32 shows how composition changes the stress-strain behavior of polyurethanes based on MDI/BDO/PPO-EO. With increasing hard segment content, the segmented polyurethanes exhibit a wide range of behavior, from a soft rubber at low hard segment content to a high modulus hard plastic at high hard segment content. The elastic modulus increases with hard segment content, which is consistent with the results of rubbery plateau modulus discussed above. Toughness as measured from the area under the stress-strain curve, however, exhibits a maximum at intermediate hard segment content where elastomeric characteristics are expected.

The above discussions only serve to illustrate some aspects of the relationships of phase separation and mechanical properties of polyurethanes and are by no means exhaustive. However, reviews by Allport and Mohajer (1973), by Noshay and McGrath (1977), by Abouzahr and Wilkes (1985), by Gibson, Vallance, and Cooper (1982) contain good summaries of what is available.

In this chapter we have seen how segmented block polymers form and phase separate. The remaining chapters discuss how these two dynamic processes can be carried out rapidly to build strong tough polymers in the RIM process. Properties of RIM produced polymers are presented in Chapters 6 and 7.

REFERENCES

Abouzahr, S.; Wilkes, G.L.; "Segmented copolymers with emphasis on segmented polyurethanes," Ch. 5 in *Processing, Structure and Properties of Block Copolymers*, Folkes, M.J. ed. Elsevier, New York, **1985**.

Alberino, L.M.; McClellan, T.R.; "The future of RIM in the United States," in *Reaction Injection Molding*, J.E. Kresta, ed., Am. Chem. Soc. Symp. Series 270, **1985**.

Allport, D.C.; Mohajer, A.A.; "Property-structure relationships in polyurethane block copolymers," in *Block Copolymers*, Allport D.C.; Jones, W.H., eds. J. Wiley, New York, **1973**.

Aleksandrova, Y.V.; Sharikov, Y.V.; Tarakanov, O.G.; "Kinetic study of the synthesis of elastic polyurethane foam," *Polym. Sci. USSR* **1970**, *11*, 2786.

Allen, G.; Egerton, P.L.; Walsh; D.J.; "Model polyurethane networks," *Polymer* **1976**, *17*, 65-71.

Andrady, A.L.; Sefcik, M.D.; "Glass transition in poly(propylene glycol) networks," *J. Polym. Sci.: Polym. Phys. Ed.* **1983**, *21*, 2453-2463.

Baker, J.W.; Bailey, D.N.; "The mechanism of the reaction of aryl isocyanates with alcohols and amines," *J. Am. Chem. Soc.* **1957**, *81*, a, Part VI, 4649; b, Part VII, 4652; c, Part VIII, 4663.

Baker, J.W.; Gaunt, J.; "The mechanism of the reaction of aryl isocyanates with alcohols and amines," *J. Am. Chem. Soc.* **1949**, *73*, 9.

Baker, J.W.; Holdsworth J.B.; "The mechanism of aromatic side-chain reaction with special reference to the polar effects of substituents," *J. Am. Chem. Soc.* **1947**, *71*, 713.

Benoit, H.; Hadziioannou, G.; "Scattering theory and properties of block copolymers with various architectures in the homogeneous bulk state,"*Macromol.* **1988**, *21*, 1449-1464.

Berry, G; Fox, T.; "The viscosity of polymers and their concentrated solutions," *Adv. Polym. Sci.* **1968**, *5*, 261.

Bidstrup, S.A.; Macosko, C.W.; "Structure and rheological changes during epoxy-amine crosslinking," *Crosslinked Epoxies*, B. Sedlacek; J. Kahovec ed. de Greyter, Berlin, 1987, 253-268.

Blake, J.W.; Anderson, R.D.; Yang, W.P.; Macosko, C.W.; "Adiabatic reactive viscometry for polyurethane reaction injection molding," *J. Rheol.* **1987**, *31*, 1236-1242.

Bonart, R.; Morbitzer, L.; Mullar, E.H.; "X-ray investigations concerning the physical structure of crosslinking in urethane elastomers III. Common structure principle for extensions with aliphatic diamines and diols," *J. Macromol. Sci.* **1974**, *9*, 447.

Borkent, G.; "Kinetics and mechanism of urethane and urea formation," *Adv. Urethane Sci. Tech.* **1974**, *3*, 1.

Born, L.; Hespe, H.; "On the physical crosslinking of amine-extended polyurethane urea elastomers: a crystallographic analysis of bis-urea from diphenyl methane-4-isocyanate and 1,4-butane diamine," *Colloid and Polym. Sci.* **1985**, *263*, 335.

Briber, R.M.; PhD Thesis, Department of Polymer Science, University of Massachusetts, 1983.

Briber, R.M.; Thomas, E.L.; "Investigation of two crystal forms in MDI/BDO-based polyurethanes," *J. Macromol. Sci. - Phys.* **1983**, *22*, 509-528.

Briber, R.M.; Thomas, E.L.; "The structure of MDI/BDO-based polyurethanes: Diffraction studies on model compounds and oriented thin films," *J. Polym. Sci: Polym. Phys. Ed.* **1985**, *23*, 1915-1932.

Broyer, E.; Macosko, C.W.; "Heat transfer and curing in polymer reaction molding," *AIChE J.*, **1976**, *22*, 268-276.

Camargo, R.E.; "Kinetics and phase segregation studies in the RIM polymerization of urethane elastomers," Ph.D. Thesis, University of Minnesota. 1984.

Camargo, R.E.; Macosko, C.W.; Tirrell, M.; Wellinghoff, S. T.; "Experimental studies of phase separation in reaction injection molded (RIM) polyurethanes," *Polym. Eng. Sci.* **1982**, *22*, 719-728.

Camargo, R.E.; Gonzalez, V.M.; Macosko, C.W.; Tirrell, M.; "Bulk polymerization kinetics by the adiabatic reactor method," *Rub. Chem. Tech.* **1983a**, *56*, 774-783.

Camargo, R.E.; Macosko, C.W.; Tirrell, M.; Wellinghof, S.T.; "Hydrogen bonding in segmented polyurethanes: band assignment for the carbonyl region," *Polym. Comm.* **1983b**, *24*, 314-315.

Camargo, R.E.; Macosko, C.W.; Tirrell, M.; Wellinghoff, S.T.; "Phase separation studies in RIM polyurethanes: catalyst and hard segment crystallinity effects," *Polymer* **1985**, *26*, 1145-1154.

Camberlin, Y; Pascault, J.P.; "Quantitative DSC evaluation of phase segregation rate in linear segmented polyurethanes and polyurethaneureas," *J. Polym. Sci.: Polym Chem. Ed.* **1983**, *21*, 415-423.

Camberlin, Y; Pascault, J.P.; "Phase segregation kinetics in segmented linear polyurethanes: Relations between equilibrium time and chain mobility and between equilibrium degree of segregation and interaction parameter," *J. Polym. Sci.: Polym. Phys. Ed.* **1984**, *22*, 1835-1844.

Casey, J.P.; Milligan, B.; Fasolka, M.J.; "Chain extender structure-activity relationships,"*J. Elast. Plast.* **1985**, *17*, 218-223.

Castro, J.M.; Macosko, C.W.; "Studies of mold filling and curing in the reaction injection molding process," *AIChE. J.* **1982**, *28*, 250-260.

Castro, J.M.; Perry, S.J.; Macosko, C.W.; "Viscosity changes during urethane polymerization with phase separation," *Polymer Comm.* **1984**, *25*, 82-87; ibid **1985**, *26*, 158.

Chambon, F; Winter, H.H.; "Stopping of crosslinking reaction in a PDMS polymer at the gel point," *Polym. Bull.* **1985**, *13*, 499-503.

Chang, A.L.; Briber, R.M.; Thomas, E.L.; Zdrahala, R.J.; Critchfield, F.E.; "Morphology study of the structure developed during the polymerization of a series of segmented polyurethanes," *Polym.* **1982**, *23*, 1060-1065.

Chang, A.L., "Processing-morphology-property relationships of poyurethanes," PhD Thesis, University of Minnesota, 1984.

Chen, Z.S.; Yang, W.P.; Macosko, C.W.; " Polyurea synthesis and properties as a function of hard segment content," *Rubber Chem. and Tech.* , **1987**, *61*, 86-99.

Christenson, C.P.; Harthcock, M.A.; Meadows, M.D.; Spell, H.L.; Howard, W.L.; Creswick, M.W.; Guerra, R.E.; Turner, R.B.; "Model MDI/Butanediol polyurethanes: Molecular structure, morphology, physical and mechanical properties," *J. Polym. Sci.: Part B: Polym. Phys.* **1986**, *24*, 1401-1439.

Coleman, M.M.; Lee, K.H.; Skvrovanek, D.J.; Painter, P.C.; "Hydrogen bonding in polymers. 4. Infrared temperature studies of a simple polyurethane," *Macromol.* **1986**, *19*, 2149.

Craven, R.L.; "Reaction rates of isocyanates with amines," Am. Chem. Soc. Sym., Atlantic City, NJ. 1957.

Critchfield, F.E.; Koleske, J.V.; Priest, D.C.; "Poly(Styrene co acrylonitrile) polyols. Modulus enhancing polyols for urethane polymers," *Rubber Chem. and Tech.* **1972**, 1467-1484.

Critchfield, F.E.; Gerkin, R.M.; "Dynamic mechanical properties of liquid reaction molded polyurethanes," *J. Elast. and Plast.* **1976**, *8*, 396-402.

Doi, M.; Edwards, S.F.; *The Theory of Polymer Dynamics*, Clarendon: Oxford 1986.

Dominguez, R.J.G.; "Amine-terminated polyether resins in RIM," *J. Cellular Plast.* **1984**, *20*, 433-436.

Dominguez, R.J.G.; Rice, D.M.; Lloyd, R.F.; "Reaction injection molded elastomer containing an internal mold release made by a two-stream system," US Patent No. 4,396,729, **1983**.

Dotson, N.A.; Galván, R.; Macosko, C.W.; "Structural development during nonlinear free-radical polymerizations," *Macromol.*, **1988**, *21*, 000.

Dusek, K.; Gallina, H.; Mikes, J.; "Features of network formation in the chain crosslinking (co)polymerization," *Polym. Bull.* **1980**, *3*, 19-25.

Dusek, K.; Ilavsky, M.; Lunak, S.; "Curing of epoxy resins. I. Statistics of curing of diexpoxides with diamines," *J. Polymer Sci. Sym.* **1975**, *53* , 29-44.

Dusek, K.; Ilavsky, M.; "Effect of dilution during network formation on cyclization and topological constraints in polyurethane networks," in *Biological and Synthetic Polymer Networks*, O. Kramer, ed. Elsevier: London, 1988.

Eisenbach, C.D.; Baumgartner, M.; Gunter, C.; "Synthesis and properties of polyurethane elastomers with monodisperse segment length distribution," *ACS Polymer Preprints*, Sept. **1985**. 7-9.

Enns, J.B.; Gillham, J.K.; "Time-temperature transformation (TTT) curve diagram: Modelling the cure behavior of thermosets," *J. App. Polym. Sci.* **1983**, *28*, 2567-2591.

Feger, C; MacKnight, W.J.; "Properties of partially cured networks. 2. The glass transition," *Macromol.* **1985**, *18*, 280-284.

Ferrarini, J.; Fowler, R.; Spatafore, N.; "Formulating reinforced RIM: Why Isocyanate choice counts," *Plast. Tech.* **1981**.

Ferry, J.D.; *Viscoelastic Properties of Polymers*, 3rd Ed., Wiley: New York, **1980**.

Flory, P.J.; *Principles of Polymer Chemistry*, Cornell Press: Ithaca, **1953**.

Galla, E.A.; Mascioli, R.L.; Bechara, I.S.; "RIM catalysis and coating interaction," *J . Elast. Plast.* **1981**, *13*, 205.

Gent, A.N.; in *Science and Technology of Rubber*, F.R. Eirich, ed., Academic Press, New York, **1978**, 419.

Gent, A.N.; Tobias, R.H.; "Threshold tear strength of some molecular networks," *ACS Symp. Ser.* **1982**, *193*, 367-376.

Gibson, P.E.; Vallance, M.A.; Copper, S.L.; "Morphology and properties of polyurethane block copolymers," Ch. 6 in *Developments in Block Copolymers - 1*, Goodman, I., ed. Appl. Sci. publishers, New York, **1982**.

Gillis, H.R.; Hurst, A.T.; Ferrarini, L.J.; Watts, A.; "Liquid MDI isocyanates: Effects of structure on properties of derived polyurethane elastomers," SPI Polyurethane Div., 6th Intern. Tech./ Marketing Conf. Nov., 1983, 375-380.

Gottlieb, M.; Macosko, C.W.; Benjamin, G.S.; Meyers, K.O.; Merrill, E.W.; "Equilibrium modulus of model polydimethylsiloxane networks," *Macromol.* **1981**, *14*, 1039-1046.

Gottlieb, M.; Macosko, C.W.; "On the supression-of-junction-fluctuations parameter in Flory's network theory," *Macromol.* **1982**, *15*, 537-541.

Gottlieb, M.; Gaylord, R.J.; "Experimental tests of entanglement models of rubber elasticity. 2. Swelling," *Macromol.* **1984**, *17*, 2024.

Graessley, W.W.; Edwards, S.F.; "Entanglement interactions in polymers and the chain contour concentration," *Polymer* **1981**, *22*, 1329.

Graessley, W.W.; "Entangled linear, branched and network polymer systems-Molecular theories," *Adv. Polym. Sci.*, **1982**, *47*, 67-118.

Grisby, R.A.; Dominguez, R.J.G.; "Polyurea RIM - A versatile high performance material," SPI meeting, November, 1985, 248-253.

Hager, S.L.; McRury, T.B.; Gerkin, R.M.; Critchfield, F.E.; "Urethane block polymers: kinetics of formation and phase development," Am. Chem. Soc. Sym. Series 172. "Urethane chemistry and applications," 1981.

Harrell, L.L.; "Segmented polyurethans. Properties as a function of segment size and distribution," *Macromol.* **1969**, *2*, 607-612.

Havranek, A.; Ilavsky, M.; Nebdal, J.; Böhm, M.; Soden, W.V.; Stoll, B.; "Viscoelastic behavior of model polyurethane networks in the main transition zone," submitted to *Colloid & Polymer Science* **1987**, *265*, 8-18.

Hickey, W.J.; Macosko, C.W.; "Rheology of randomly branched polydimethyl siloxanes," *ACS Polymer Preprints* **1981**, *22*, 379-380.

Ilavsky, M.; Dusek, K.; "The structure and elasticity of polyurethane networks: 1. Model networks of poly(oxypropylene) triols and diisocyanate," *Polymer* **1983**, *24*, 981-990.

Koberstein, J.T.; Russell, T.P.; "Simultaneous SAXS-DSC study of multiple endothermic behavior in polyether-based polyurethane block copolymers," *Macromol.* **1986**, *19*, 714-720.

Kolodziej, P.; Yang, W.P.; Macosko, C.W.; Wellinghoff, S.T.: "Impingement mixing and its effect on the microstructure of RIM polyurethanes," *J. Polym. Sci.* **1986**, *24*, 2359.

König, K.; Dietrich, M.; (to Bayer AG), US Patent No. 4,089,835 **1978**.

Kovacs, A.J.; *Fortschr. Hochpolym. Forsch.* **1963**, *3*, 394.

Kuryla, W.C.; Critchfield, F.E.; Platt, L.W.; Stamberger, P.; "Polymer/Polyols, a new class of polyurethane intermediates," *J. Cell. Plast.* **1966**, 84-96.

Kwei, T.K.; "Phase separation in segmented polyurethanes," *J. App. Polym Sci.* **1982**, *27*, 2891-2899.

Landin, D.T.; Macosko, C.W.; "Cyclization and reduced reactivity of pendant vinyls during the copolymerization of methyl methacrylate and ethylene glycol dimethacrylate," *Macromol.*, **1988**, *21*, 846-851.

Landin, D.T.; Macosko, C.W.; "Rheological changes during the copolymerization of vinyl and divinyl monomers," in "Characterization of highly crosslinked polymers," Labana, S.S., Dickie, R.A., Eds. Am. Chem. Soc. Symp. Series 243. Washington, D.C., 1984.

Lawler, L.F.; Gill, W.A.; "Thermal and pressure phenomena during reaction injection molding," Am. Inst. Chem. Eng. National Meeting, Detroit, August, 1981.

Leibler, L.; "Theory of microphase separation in block copolymers," *Macromol.* **1980**, *13*, 1602-1617.

Leung, L.M.; Koberstein, J.T.; "DSC annealing study of microphase separation and multiple endothermic behavior in polyether-based polyurethane block copolymers," *Macromol.* **1986**, *19*, 706-713.

Lipshitz, S.D.; Macosko, C.W.; "Rheological changes during a urethane network polymerization," *Polym. Eng. Sci.* **1976**, *16* , 803-810.

Lipshitz, S.D.; Macosko, C.W.; "Kinetics and energetics of a fast polyurethane cure," *J. App. Polym. Sci.* **1977**, *21*, 2029-2039.

Lopez-Serrano, F.; Castro, J. M.; Macosko, C.W.; Tirrell, M.; "Recursive approach to copolymerization statistics," *Polymer.* **1980**, *21*, 263-273.

Lovering, E.G.; Laidler, K.J.; "Thermochemical studies of some alcohol isocyanate reactions," *Can. J. of Chem.* **1960**, *40*, 26.

Lyman, D.J.; "Polyurethanes," *Rev. Macromol. Chem.* **1966**, *1*, 191.

Lyman, D.J.; "Polyurethanes, The chemistry of the diisocyanate-diol reaction," in *Step-growth polymerizations*, Marcel Dekker: New York, 1972.

Macosko, C.W.; Miller, D.R.; "A new derivation of average molecular weights of nonlinear polymers," *Macromol..* **1976**, *9*, 199-206.

Manas-Zloczower, I.; Macosko, C.W. "Moldability diagrams for reaction injection molding of a polyurethane crosslinking system," *Polm. Eng. Sci.*, **1988**, *28*, 000.

Miller, D.R.; Macosko, C.W.; "A new derivation of post gel properties of network polymers," *Macromol.* **1976**, *9*, 206-211.

Miller, D.R.; Macosko, C.W.; "Average property relations for nonlinear polymerization with unequal reactivity," *Macromol.* **1978**, *11*, 656-662.

Miller, D.R.; Macosko, C.W.; "Substitution effects in property relations for stepwise polyfunctional polymerization," *Macromol..* **1980**, *13*, 1063-1069.

Miller, D.R.; Macosko, C.W.; "Calculation of average network parameters using combined kinetic and Markovian analysis," in *Biological and Synthetic Polymer Networks*, O. Kramer, ed., Elsevier: London, 1988, 219-232.

Miller, D.R.; Valles, E.M.; Macosko, C.W.; "Calculation of molecular parameters for stepwise polyfunctional polymerization," *Polym. Sci. Eng.* **1979**, *19*, 272-283.

Nishimura, H.; Kojima, H.; Yarita, T.; Noshiro, M.; "Phase structure of polyetherpolyol-4,4'-diphenylmethane diisocyanate-based reaction injection molded (RIM) polyurethanes," *Polym. Eng. and Sci.* **1986**, *26*, 585-592.

Nissen, D.; Markovs, R.A.; "Aromatic diamines as chain extenders in RIM urethane elastomers," *J. Elast. Plast.* **1983**, *15*, 71-78.

Ng, H.N.; Allegrezza, A.E.; Seymour, R.W.; Cooper, S.L.; "Effect of segment size and polydispersity on the properties of polyurethane block polymers," *Polymer* **1973**, *14*, 255.

Noshay, A.; McGrath, J.E.; *Block copolymers: Overview and critical survey*, Academic Press: New York, **1977**.

Ogden, R.W.; "Recent advances in the phenomenological theory of rubber elasticity," *Rubber Chem. Tech.* **1986**, *59*, 361-383.

Pannone, M.C.; "Kinetics of the reactions of aromatic isocyanates with alcohols and amines," M.S. Thesis, University of Minnesota. 1985.

Pannone, M.C.; Macosko, C.W.; " Reaction kinetics of a polyurea reaction injection molding system," *Polym. Eng. and Sci.* **1988**, *28*, 660-669.

Pannone, M.C.; Macosko, C.W.; " Kinetics of isocyanate amine reactions," *J. Appl. Polym. Sci.* **1987**, *34*, 2409-2432.

Pearson, D.S.; Graessley, W.W.; "The structure of rubber networks with multifunctional junctions," *Macromol.* **1978**, *11*, 528-533.

Perry, S.J.; "Rheology of polyurethane reaction injection molding systems," M.S. Thesis, University of Minnesota. 1982.

Perry, S.J.; Castro, J.M.; Macosko, C.W.; "A viscometer for fast polymerizing systems," *J. Rheol.* **1985**, *29*, 19-36.

Phillips, B.A.; Taylor, R.P.; "Polyurea dispersions for RIM applications," *Rubber Chem. and Tech.* **1979**, *52*, 864-870.

Prime, R.B.; "Thermosets" in *Thermal Characterization of Polymeric Materials*, E.A. Turi, ed. Academic Press: New York **1981**, 435-571.

Reegen, S.L.; Frisch, K.C.; "Catalysis in isocyanate reactions," *Adv. Urethane Sci. Technol.* **1971**, *1*, 1.

Reischl, A.; Jabs, G.; Dietrich, W.; Carlos, A. (to Bayer AG); US Patents 4,092,275 and 4,093,569, **1978**.

Richter, E.B.; Macosko, C.W.; "Kinetics of fast (RIM) urethane polymerization," *Polym. Eng. Sci.* **1978**, *18*, 1012-1018.

Richter, E.B.; Macosko, C.W.; "Viscosity changes during isothermal and adiabatic urethane network polymerization," *Polym. Eng. Sci.* **1980**, *20*, 921-924.

Rietsch, F.; Davelosse, D.; Froelich, D.; "Glass transition temperature of ideal polymeric networks," *Polymer* **1976**, *17*, 859-862.

Robins, J.; Edwards, B.H.; Tokach, S.K.; "A caloric investigation of the reaction between phenyl isocyanate and ethylene glycol catalyzed by dibutyltin compounds," *Adv. Urethane Sci. Technol.* **1984**, *9*, 65.

Ryan, A.J. PhD Thesis, University of Manchester, Inst. of Sci. and Tech., 1988.

Ryan, A.J.; Stanford, J.L.; Still, R.H. "Calculation of the solubility parameters of typical structural units present in segmented polyurethanes. Poly(urethane-ureas) and polyureas formed by reaction injection moulding," *Polm. Commun.* **1988**, *29*, 196-198.

Sarmoria, C.; Valles, E.M.; Miller, D.R.; "Ring-chain competition kinetic models for linear and non-linear step-reaction copolymerizations," *Makromol. Symp.* **1986**, *2*, 69-87.

Saunders, J.H.; Frisch, K.C. *Polyurethanes chemistry and technology*, vols. 1 and 2, Wiley-Interscience: New York, 1962.

Scholtens, B.J.R.; "Linear Thermoviscoelasticity and characterization of noncrystalline EPDM rubber networks," *J. Polym. Sci.: Phys. Ed.* **1984**, *22*, 317-344.

Scott, R.L. ; "The Thermodynamics of high polymer solutions. V. Phase equilibria in the ternary system: polymer 1-polymer 2-solvent," *J. of Chem. Phys.* **1949**, *17*, 279-284.

Seefried Jr., C.G.; Koleske, J.V.; Critchfield, F.E.; "Thermoplastic urethane elastomers. I. Effects of soft-segment variations," *J. Appl. Polym. Sci.* **1975**, *19*, 2493.

Smith, T.L.; *Rubber Chem. Tech.* **1978**, *51*, 224.

Smith, T.L.; "Elastic modulus and other mechanical properties of single-phase polyurethane elastomers," *ACS Symp. Ser.* **1982**, *193*, 419-437.

Speckhard, T.A.; Cooper, S.L.; "Ultimate tensile properties of segmented polyurethane elastomers: Factors leading to reduced properties for polyurethanes based on nonpolar soft segments," *Rubber Chem. and Tech.* **1986**, *59*, 405-431.

Steinle, E.C.; Critchfield, F.E.; Castro, J.M.; Macosko, C.W.; "Kinetics and conversion monitoring in a RIM thermoplastic polyurethane system," *J. Appl. Polym. Sci.* **1980**, *25*, 2317-2329.

Stepto, R.F.T.; in *Developments in polymerization. 3*, Haward, R.N. ed; Applied Science Publishers: Barking, 1982.

Sung, P.-H.; Mark, J.E.; "Model polyurethane elastomers prepared from noncrystallizable poly(propylene oxide) chains," *J. Polym. Sci.* **1981**, *19*, 507-515.

Taylor, R.P.; Dewhurst, J.E.; Abouzahr, A.M.; "Process and composition for the production of polyure-thane elastomer moldings," US Patent No. 4,442,235, **1984**.

Vallés, E.M.; Macosko, C.W.; "Structure and viscosity of polydimethyl siloxanes with random branches," *Macromol.* **1979**a, *12*, 521-526.

Vallés, E.M.; Macosko, C.W.; "Properties of networks formed by end linking poly(dimethylsiloxane)," *Macromol.* **1979**b. *12*, 673-679.

Vespoli, N.P.; Alberino, L.M.; Peterson, A.A.; Ewen, J.H.; "Mold filling studies of polyurea RIM systems," *J. Elast. Plast.* **1986**, *18*, 159-175.

Vespoli, N.P.; Alberino, L.M.; "Comparison of rheological measurements during mold filling for glycol and amine extended urethane RIM systems," *J. Elast. Plast.* **1985**b, *17*, 173-182.

Vespoli, N.P.; Alberino, L.M. "Computer modeling of heat transfer processes and reaction kinetics of urethane-modified isocyanurate RIM systems," *Polym. Process Eng.*, **1985**, *3*, 127-148.

Wilkes, G.L.; Widnauer, R.; "Kinetic behavior of the thermal and mechanical properties of segmented urethanes," *J. Appl. Phys.* **1975**, *46*, 4148-4152.

Wilkes, G.L.; Emerson, J. A.; "Time dependence on small-angle x-ray measurements on segmented polyurethanes following thermal treatment," *J. Appl. Physics* **1976**, *47*, 4261-4264.

Willkomm, W.R.; Chen, Z.S.; Macosko, C.W.; Gobran, D.A.; Thomas, E.L. "Properties and phase separation of RIM and solution polymerized polyureas as a function of hard block content," *Polym. Eng. Sci.*, **1988**, *28*, 000.

Williams, R.J.J.; "Curing of thermosets," in *Developments in Plastics Technology-2*, A. Whelan, J.L. Craft, ed., Elsevier Applied Science: London, 1986. in press.

Woods, G.; *Flexible polyurethane foams: Chemistry and technology*, Applied Science Publishers: London, 1982.

Wright, P.; Cumming, A.P.C.; *Solid polyurethane elastomers*, MacLaren and Sons: London, 1969.

Yang, W.P.; "Dynamics of phase separation and its effect on polyureathane structure and properties" Ph.D. Thesis, University of Minnesota. 1987.

Yang, W.P.; Macosko, C.W.; Wellinghoff, S.T.; "Thermal degradation of urethanes based on 4,4'-diphenylmethane diisocyanate and 1,4-butanediol (MDI/BDO)," *Polymer* **1986**, *27*, 1235-1240.

Yang, W.P.; Macosko, C.W.; "Phase separation during fast (RIM) polyurethane polymerization, *Makromol. Symp.*, **1989**, *25*, 000.

Yeakey, E.L.; "Process of preparing polyoxyalkylene polyamines," US Patent No. 3,654,370, **1972**.

Zdrahala, R.J.; Gerkin, R.M.; Hager, S.L.; Critchfield, F.E.; "Polyether-based thermoplastic polyurethanes. I. Effect of the hard-segment content," *J. App. Polym. Sci.* **1979**, *24*, 2041-2050.

Zdrahala, R.J.; Critchfield, F.E.; "RIM urethanes structure property relationships for linear polymers," in *Reaction Injection Molding and Fast Polymerization Reactions*, J.E. Kresta, ed. Plenum Press, New York, **1981**, 55-62.

3

MATERIAL DELIVERY

Chapter 1 gave a brief overview of the RIM process. In this chapter we will look in more detail at the first three unit operations shown in Figure 1.2: component supply, low pressure recirculation or conditioning and high pressure metering. Subsequent chapters will deal with the mixing, filling and curing steps. This chapter has two goals. The first and most important is to identify the main variables that must be controlled before the mixing step, i.e. before polymerization begins: temperature, component dispersion, gas nucleation, pressure, reactant flow rates and ratio. The second goal is to describe how this is accomplished with current equipment. The chapter will concentrate on general equipment principles rather than attempt to list all the manufacturers and delineate their differences. Books by Sweeney (1979, 1987), Becker (1979) and Oertel (Ch. 4, 1985), as well as some articles cited below, give more details on particular RIM machinery.

3.1 Process Description

Figure 3.1 shows a typical RIM machine and Figure 3.2 gives a schematic diagram. Its major components are 1) day tanks with low pressure recirculation pumps, 2) high pressure injection pistons and 3) mixhead. Since each machine can service several molds, the mold with its opening and closing clamp is usually treated as a separate unit (Chapters 5 and 6). Nearly all machines are designed for two components and the design of each component side is very similar.

Most of the machine time is spent in low pressure recycle. Reactants recirculate from their day tanks, through the metering cylinders, mixhead and back to the tanks through heat exchangers. This insures temperature uniformity at the beginning of a shot (injection into the mold) and aids in dispersion.

About 15 s before a shot is needed the cycle is initialized: the mold is closed and the metering pistons are allowed to move backwards, filling up with reactants (Figure 3.3a). About 5 s before a shot the pistons are pressurized (Figure 3.3b). This provides time for the flexible hoses which feed the mixhead to expand and for reactants to accelerate to the high velocity needed for impingement mixing.

When both reactant streams are recycling at the proper velocity, the mixhead is rapidly opened and the mold is filled in 1 or 2 s. (Figure 3.3c). The mixhead quickly closes and the reactants again recycle under high pressure to empty the metering cylinders. Then the machine switches back to low pressure recycle again.

Meanwhile polymerization or curing is occurring inside the closed mold, building up enough stiffness and strength such that the mold can be opened and the part removed. Times for these various steps for an automobile fascia part are depicted in the RIM cycle of Figures 3.4 and 3.5. Note that less than 15% (about 10 s) of the machine cycle is under high pressure and only 1/3 of the mold cycle is required for polymerization. Recent developments with polyureas have made 30s demold times possible. The relatively long part removal period reflects time for the press to tilt outward, permitting easier removal of large parts.

Since during a molding cycle only a small proportion of time is required at high pressure, one machine can service more than one mold. This is shown schematically in Figure 3.6. Multiple mold stations are often used in high volume production and sometimes in applications where mold cost is low and cure time longer as in furniture parts. For automotive production the current trend is to have a separate high pressure unit for each mold but day tanks which serve three or four stations.

Figure 3.1   A typical RIM machine, without mold (from Battenfield product literature, 1986).

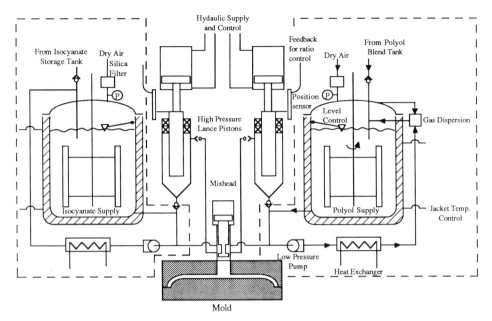

Figure 3.2   Schematic of a typical RIM machine. The machine can be divided into 3 basic parts: 1) low pressure recirculation or conditioning (bounded by the dotted lines); 2) high pressure metering; and 3) the impingement mixhead. The mold is usually considered separately. The figure shows the machine in low pressure recycle mode.

In addition to RIM machines and molds, a complete reaction injection molding plant requires large storage tanks to receive tank truck or rail car deliveries of components. All tanks must be inert to chemical attack. Some formulations can be stored in carbon steel tanks, others require stainless steel. A phenolic spray coating can also be used to protect carbon steel. The tanks are blanketed with dry air or nitrogen to prevent moisture absorption. Some temperature control is needed to prevent freezing of isocyanates. Temperature homogenization is assisted by slow recirculation. Typically storage tank contents are recirculated every 6 hours.

The liquified isocyanates are typically pumped directly to the supply tank of each RIM machine. The polyol side consists of several components which are combined in a blend tank. This is shown schematically in Figure 3.6 along with the machine and mold layout in a typical plant (Sweeney, 1979; Oertel, Ch. 4, 1985; Bayer, 1985).

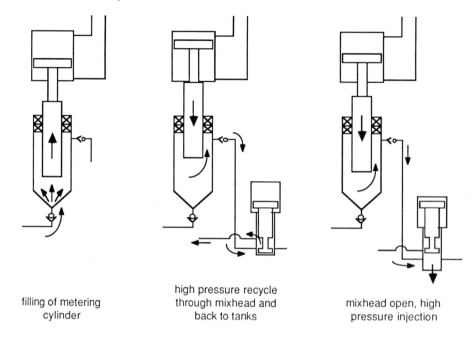

| filling of metering cylinder | high pressure recycle through mixhead and back to tanks | mixhead open, high pressure injection |

Figure 3.3   Schematic of high pressure piston and mixhead during various stages: a) low pressure filling of the metering cylinders, b) high pressure recycle through the mixhead and c) mixhead opening followed by high pressure injection into the mold.

3.2 Low Pressure Conditioning

During low pressure recirculation several variables must be controlled. Liquid level in the supply tanks is maintained by standard float controllers. These call for intermittent refilling from the storage tanks, typically every one or two injections.

Temperature must be maintained to ±2°C to control reactivity and viscosity for impingement mixing. For urethane systems this temperature is not much above ambient, typically 30-50°C. Since high pressure recycling can dissipate significant energy, heat removal is frequently necessary through water regulated heat exchangers. The supply tanks are jacketed with tempered water or oil for high temperature applications. Thermocouples are located in the supply tanks, near the mixhead and in the return lines. Their readings are displayed and used by the controller to regulate the heat exchanger (Peters, 1983; Schneider, 1984; Bayer, 1985).

For nylon and epoxy formulations the reactants must be maintained at higher temperatures. For example, the melting point of caprolactam, the monomer for nylon 6, is

67°C, thus all lines must be heat traced to avoid cold spots and freezing. The metering cylinders are sometimes placed inside heated cabinets to maintain temperature (Hall, 1985).

Low pressure recirculation is maintained by simple gear pumps. As indicated in Figure 3.2 valves can be used so that the recycle bypasses the metering cylinders and the mixhead. When mineral fillers are present an eccentric screw pump, such as a Moyno type, is required. For unfilled reactants in-line filters prevent large particles from jamming mixhead orifices; 0.15mm mesh is typical. Recirculation helps to keep fillers and immiscible liquids dispersed but tank stirrers (see Figure 3.2) are also used.

Another important function of the recirculation line is to disperse dry air or nitrogen into one or both of the reactants. Dry air or nitrogen serves to protect the reactants from water contamination. In the case of some non-urethane chemistry, (see Chapter 7) nitrogen is required to protect the reactants from oxygen. It also helps to maintain positive pressure

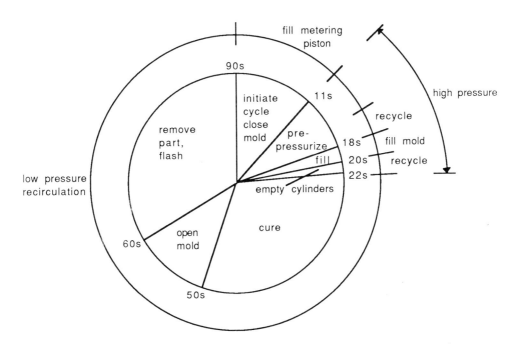

Figure 3.4   Typical RIM cycle for a 5 kg urea-urethane automobile fascia similar to Formulation-5 Table 2.4 (adapted from Cekoric, Taylor and Barrickman, 1983). Inner circle is for the mold and mixhead; outer circle for the metering system.

(2-3 atm) on the recirculation pumps. The dispersed gas bubbles provide a means for packing the mold during the curing stage. During recycle, dry air is finely dispersed into the polyol through a rotating injector or venturi in the return line. The density of the polyol is reduced to 0.6-0.8. These bubbles collapse in the high pressure metering cylinders but reappear during decompression in the mixhead. They expand with the reaction exotherm to compensate for shrinkage which occurs during polymerization. Mold packing is discussed further in Chapter 6.

Control of gas content in the reactants is necessary for maintaining correct mold packing. An intermittent check can be made by drawing off a sample into an isolated section of pipe and allowing it to expand against a piston under low pressure (Becker, 1979, Ch.8). But continuous on-line density measurements are preferable. Both gamma radiation and oscillating tube detectors have been successfully placed in the recycle line (Moser, 1983; Peters, 1983). Control of the initial bubble size distribution through injector design and surfactants is also important for avoiding large bubbles in the final part.

3.3 High Pressure Metering

The low pressure conditioning portion of a RIM machine supplies a well dispersed mixture of components at the proper temperature and gas loading level to the high pressure metering unit. The metering unit must deliver the reactants to the mixhead at 50-250 atm, at

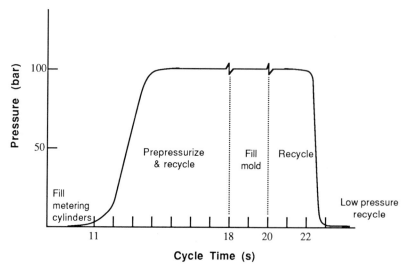

Figure 3.5   Pressure versus time at the inlet to the mixhead during a typical RIM cycle (adapted from Oertel, Ch. 4, 1985). Transients in the high pressure signal occur when the mixhead slides from recycle to inject mode and vice versa.

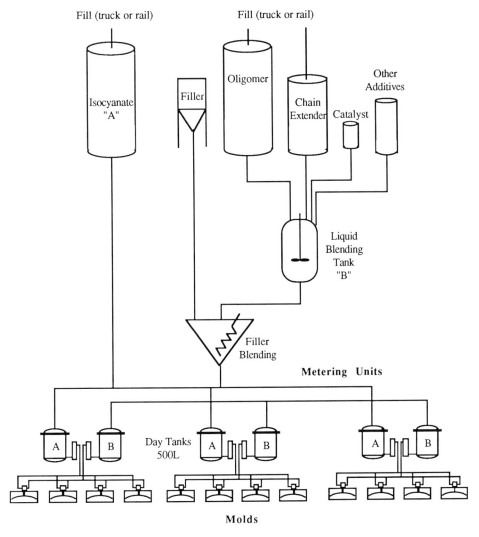

Figure 3.6 Schematic of a RIM plant. Large storage tanks are filled by truck or rail deliveries. Compounding several components is done in a blend tank which in turn feeds the day tanks on each RIM machine. One machine can each service up to 4 molds.

the desired flow rate and at precisely controlled, $< \pm 1\%$, flow ratio. Two types of high pressure metering pumps are used: multistroke and single stroke.

Multistroke high pressure piston pumps were used on the first RIM machines and are still the pump of choice for high pressure foam production. There are several designs: radial piston, axial (or bent axis) and vertical (or in-line) pistons which are described in detail elsewhere (Becker, 1979; Sweeney, 1979; Oertel, Ch. 4, 1985). All designs convert the rotation of a constant speed motor into a sequential motion of several pistons; six or more give an output smooth enough for impingement mixing. Volumetric flow is increased by increasing the stroke in the case of the radial or axial pumps. In the case of the vertical pumps flow is increased by covering the inlet port earlier in the stroke. Typically two pumps are mounted on one motor shaft. This insures very precise ratio control. The main errors are from leakage. Flow rate can be monitored on-line by induction counting the rotations of a gear enclosed in the line.

The major disadvantage of these pumps for RIM is that the small valves and tight metal-to-metal seals which make them precise over a wide viscosity, flow rate, and pressure range make it impossible to use hard fillers like glass or minerals. They also are limited to viscosities below 10 poise and are not available with the high flow rates, ≥ 2 L/s, needed for very fast reacting systems.

Multistroke pumps, however, are clearly advantageous for large parts like rigid foam insulation or furniture. If chemical reactivity can be controlled, a small machine can fill a very large mold. Another advantage is that in some cases the same continuous pumps can be used for low pressure recirculation by running the drive motor at low speed.

Axial piston pumps are the main choice for unfilled RIM and high pressure foam machines. Vertical piston pumps were developed for diesel fuel injection. They are more expensive than the axial type but are more precise at low flow rate (0.1 - 1.5 L/min) and low viscosity. They are used for small RIM machines and for dosing third components like colorant or mold release (Moser, 1983).

Single stroke or **displacement** pumps are the main choice for RIM because of their ability to handle fillers and higher flow rate. There are two types: piston and lance. They are quite similar except for the location of the high pressure seals. In the piston displacement pump the seals are mounted on the moving piston and wipe the walls clean of liquid, much like a standard hydraulic cylinder. In the lance or plunger type displacement pump, shown schematically in Figures 3.2 and 3, the seals are fixed in the cylinder at one end. Fluid is displaced out of the cylinder by the motion of the smooth walled lance. The fixed seals function better with abrasive fillers and thus are preferred for RIM. However, they always leave some reactant on the cylinder walls and thus have more dead volume than the piston type.

The piston displacement pump can be driven directly at the rear of the piston by hydraulic fluid. However, the hydraulic oil can become contaminated by leakage and most machines attach a separate hydraulic cylinder as indicated in Figure 3.2. This cylinder can be controlled by one of the continuous pumps described above. Thus a displacement cylinder can be added to convert a multistroke machine to one for filled reactants.

Today most RIM machines are designed for filled reactants and high flow rates; the majority use lance pistons driven by hydraulic cylinders. These drive cylinders are powered by a hydraulic pump and accumulator. As indicated in Figure 3.2, to control ratio most machines monitor motion of the lance and regulate hydraulic oil flow rate to each cylinder with fast responding servo valves. The results of Coates, Sivakumar and Johnson (1986) shown in Table 3.1 demonstrate that with feedback control stoichiometric ratio can be controlled to less than 1%. Commercial RIM equipment manufacturers report similar values (Sneller, 1986). Feedback control can also be based on measurement of the flow rate of the oil to the hydraulic drive cylinders (Brownbill,1983). Current machines use microprocessors to control the valve sequencing during a cycle.

Although nearly all RIM machines are built for two components, there is some activity in three or four component metering. A small third stream can be used to add colorants, catalysts or mold release agents. Nguyen and Suh (1984) and Coates, et al. (1987) have used multiple stream machines to make interpenetrating network polymers. Coates et al. have pointed out that such a machine could even be used to change composition during a shot, producing, for example, a product with an elastomeric surface and a rigid core. If a machine with four lance pistons is operated with only two components its output can be doubled. With this approach 10 L/s flow rates are possible (*Plast. Eng.*, January 1988). If two pistons alternate very large shots and even continuous output are possible.

Table 3.1   Lance Piston RIM Machine Volumetric Flow Rate Errors*

	maximum	rms	total
open loop	+6.5, -2.5%	2.5%	4.1%
feed back control	+1.7, -1.5	0.6	0.08

* based on a 2.2s shot at 1 L/s and 1.225 ratio, 8 wt% glass in polyol (Coates, Sivakumar and Johnson, 1986).

3.4 Lab Scale Machines

Surprisingly, small scale RIM equipment is not readily available. Since the main application for RIM has been large parts, most commercial machines are designed for 100 L or more capacity. Even if they are capable of small shots, the volume of reactants and the time needed to flush out tanks, pumps and lines makes it difficult to use these large machines for research into new RIM chemistry or even for development work on current formulations.

Early formulations were developed by fast hand mixing using reduced catalyst levels in laboratory glassware. This approach fails with the present high speed RIM systems, particularly the ureas. Thus a number of small volume machines have been designed (Oertel, Ch. 4, 1985; Mikkelsen and Macosko, 1988).

The simplest is a fixed ratio type. Like a pair of syringes, two small hydraulic cylinders can be driven simultaneously by a third power cylinder. Cylinder diameter controls stoichiometric ratio and, by using a large power cylinder, compressed gas can be used as the power source. The arrangement of the cylinders is similar to that shown in Figure 3.7 but with the lever arm replaced by a fixed bar. Richter and Macosko (1978) were able to make samples from 100 ml of each reactant with such a design. Components must be well dispersed externally and then loaded manually for each shot. With lance displacement pumps abrasive fillers have been used (Mikkelsen and Macosko, 1988). Temperature is controlled by heating tape or putting the unit into an oven. Maintaining a large difference in temperature between the two reactants on such a small machine can be difficult.

To achieve variable ratio without the cost and size required for feed back control several workers (Lipshitz, 1974; Ishida and Scott, 1986) and at least one commercial supplier (Accuratio Systems Inc., see Sweeney, 1979) have used ball screws to drive the displacement pumps. A gear between the two screw drives maintains ratio. This gear is in turn driven by a variable speed motor. With different gears the stoichiometric ratio can be changed, not continuously but usually with satisfactory increments. An approach which provides infinite resolution of ratio is to use two motors (Ishida and Scott, 1986). The ball screw design appears to be more limited in maximum pressure and flow rate than the hydraulic or air powered systems.

Lee and Macosko (1979) used the lever arm system shown in Figure 3.7 to control ratio. This provides continuous variation of ratio over a limited range. For example, two 2 inch diameter cylinders cover the range 2.0:1 to 1.26:1; while a 2 inch and a 1-1/2 inch cylinder give 0.71:1 to 1.12:1 or 2.24:1 to 3.55:1 if their positions are reversed. Flow rates up to 100 ml/s and pressures to 150 bar can be developed using 10 bar nitrogen in the power cylinder. Lee and Macosko (1979) and Sandell (1984) give equations for relating flow rate to pressure drop. 400 - 500 ml of each reactant is sufficient to mold RIM plaques. This design has been duplicated in a number of laboratories and a large variety of materials have been molded.

Material conditioning has not been optimized with this design. Recirculation is accomplished by low pressure movement of the metering cylinders. There has yet been no provision for gas injection and thus no mold packing control. Parts have surface sink

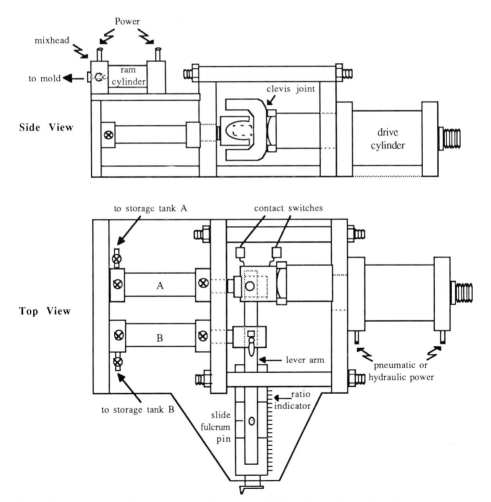

Figure 3.7   Schematic of level variable ratio lab scale RIM machine (adapted from Lee and Macosko, 1979; and Macosko and Lee, 1980).

marks. However, mechanical properties of typical high density parts are quite comparable to those for plaques produced with production scale machines. One example with a commercial RIM formulation is shown in Table 3.2. (Kolodziej, 1980).

A larger lab scale RIM machine can be constructed from small vertical, multistroke piston pump (Charbonneaux, 1988). It can provide easier recirculation but requires a minimum of 6 L of each reactant and can not process filled liquids.

TABLE 3.2 Comparison of Mini-RIM Plaque Mold Properties to Typical Properties of the Polyurethane RIM 2600 System (Union Carbide)*

RIM 2600	Typical	Mini-RIM
Process Conditions		
Polyol Temp. (°C)	55	60
143 - L Temp. (°C)	23	30
Mold. Temp. (°C)	71	100
Post Cure (hr/°C)	1/121	1/100
Physical Properties		
Hardness, Shore D	58	55
100% Modulus (MPa)	15.9	17.9 (2590 psi)
Tensile Strength (MPa)	23.4	23.6 (3430 psi)
Ultimate Elongation (%)	235	200
Flex Modulus (MPa)		
-29°C	480	670
24°C	180	190 (28,000 psi)
70°C	100	90

*Kolodziej, 1980

REFERENCES

Bayer "RIM technology, concepts for a PUR-RIM line," PU-Applications Research Department Report, Leverkusen: April 1985.

Becker, W.E., ed. *Reaction Injection Molding*, Van Nostrand Reinhold: New York, 1979.

Brownbill, D. "Process control comes to RIM," *Mod. Plast., Int. Ed.*, **1983**, *60 no. 6*, 22-24.

Cekoric, M.E.; Taylor R.P.; Barrickman,C.E. "Internal mold release, the next step forward in RIM," SAE Tech. Paper #830488, Detroit, March 1983.

Charbonneaux, T.G. "Continuous reactive processing of crosslinkable monomers," PhD Thesis, University of Minnesota, 1988.

Coates, P.D.; Sivakumar. A.I.; Johnson, A.F. "RRIM: Machine and process measurements," *Polym. Process. Eng.* **1986**, *3*, 219-234.

Coates, P.D.; Johnson, A.F.; Armitage, P.D.; Hynds, J.; Leadbitter, J. "Novel reinforced RIM processing: computer controlled multiple stream RRIM production of interpenetrating networks and polyurethanes," *Polym. Eng. Sci.* **1987**, *27*, 1209-1215.

Hall, C.M. "RIM flexes its muscles in seeking out new markets," *Plastics Eng.*, **1985**, June, 39.

Ishida, H.; Scott, C. "Fast polymerization and crystallization kinetic studies of nylon-6 by combined use of computerized micro-RIM machine and FT-IR," *J. Polym. Eng.*, **1986**, *6*, 201.

Kolodziej, P. "The effect of impingement mixing on the morphology of RIM polyurethanes," MS Thesis, University of Minnesota, 1980.

Lee, L.J.; Macosko, C.W. "Design and characterization of a small reaction injection molding machine," *Soc. Plast. Eng. Tech. Papers* **1979**, *24*, 151-154.

Lipshitz, S.D., "Laminar tube flow with a thermosetting urethane polymerization," PhD Thesis, University of Minnesota, 1976.

Macosko, C.W.; Lee, L.J. "Reaction injection molding machine," US Patent 4,189,070 (1980).

Mikkelsen, K.J.; Macosko, C.W. "Characterization of laboratory RIM machines," *J. Cellular Plast.* **1988**, *24*, 000.

Moser, K. "RIM and RRIM technology," Kunststoffe **1983**, *73* , 583-587; ibid, 764-767.

Nguyen, L.T.; Suh, N.P. "Reaction injection molding of interpenetrating polymer networks," Riew, C.K.; Gillham, J.K.; eds. Am. Chem. Soc. Symp. Series 208, Washington D.C. 1984.

Oertel, G., ed. *Polyurethane Handbook*, Hanser: Munich, 1985.

Peters, G.M. "Exacting RIM process controls help yield higher-quality plastic parts," *Plast. Eng.* **1983**, November, 33-35.

Richter, E.B.; Macosko, C.W. "Kinetics of fast (RIM) urethane polymerization," *Poly. Eng. Sci.* **1978**, *18*, 1012-1018.

Sandell, D.J., "Experimental methods for studying impingement mixing," MS Thesis, University of Minnesota, 1983.

Schneider, P.S. "The basics of today's RIM controls," *Plast. Tech.*, **1984**, March, 67.

Sneller, J.A. "Detroit's plastics planning sparks rise in RIM automation," *Mod. Plast.*, **1986**, *63 no. 2*, 55-58.

Sweeney, F.M., *Introduction to Reaction Injection Molding*, Technomic: Westport, CT, 1979.

Sweeney, F.M., *Reaction Injection Molding Machinery and Processes*, Marcel Dekker: New York, 1987.

4

MIXING

The equipment described in the previous chapter is all designed to prepare reactive components for delivery to the mixhead. Now let us examine that critical part of the process.

High jet velocity impingement mixing is often called the heart of RIM. It is what differentiates RIM from other reaction molding processes like thermoset injection molding or sheet molding. The latter are activated by heat transfer from hot mold walls. In RIM, polymerization is activated by contact between reactants in the impingement mixing step. This frees RIM from heat transfer limitations making possible extremely fast molding of large, thick parts.

Despite its central role to RIM and considerable research, impingement mixing is still not well understood. Thus let us first look at some of the most common commercial mixheads, picking out the key features that seem to make them successful. Then we will try to dig deeper to learn what is going on inside the mixing chamber.

4.1 Mixhead Design

RIM is a batch operation. There needs to be a means to start and stop the mixing and to keep the mixing chamber clear of polymer. Low pressure machines for urethane foam and for dispensing other reactive systems use a flow through, stirring mixer. Valves on each stream start and stop the flow and a solvent or air is used to flush out the reactants before they can gel. This is too messy and too slow for RIM.

It has been known for some time that impingement of two streams into a pipe makes a simple and effective continuous mixer. For example Narayan showed that wall vs center concentration differences could be eliminated in only 2 diameters in a "T" mixer vs 250 diameters for a concentric cylinder or jet mixer (Tucker and Suh, 1980).

Boden, Schulte and Wirtz (Oertel, 1985) have reviewed the evolution of impingement mixhead design for RIM. Today all RIM mixheads employ the following features:

- recirculation of reactants near the mixing chamber to maintain temperature and dispersion and to permit rapid start up of mixing
- rapid opening valve(s) to control stoichiometry from start up of mixing
- inlet nozzles which accelerate reactant jets to high velocity

- chamber where streams impinge and initial mixing occurs
- some means to insure that the chamber is filled rapidly and proper impingement occurs from the beginning of the shot
- clean out piston to push all reactants out of the mixing chamber

The majority of RIM mixheads contain two inlets but larger heads can have three or four. Thus multicomponent mixing is possible. In practice a third stream has been used mainly for a small flow rate of additives like color pigments or mold release agents.

The most common mixhead design is shown in Figure 4.1 (Keuerleber and Pahl, 1970; Pahl and Schlüter, 1971; Wingard and Leidal, 1978). In the recirculation mode reactants flow through grooves in the clean out piston back into the supply tanks. When the piston is retracted reactants can flow into the mixing chamber. When the piston moves forward again injection is interrupted, recirculation reestablished and material in the chamber is pushed into the mold. The effective velocity at the inlet nozzle can be increased by decreasing the orifice diameter or by advancing the screws into the nozzles which restrict flow.

A variation of this design is shown in Figure 4.2 where the recirculation grooves are replaced by holes through the clean out piston. Sealing against cross flow between the holes may be easier when compared to the long grooves of Figure 4.1. This·design has been used for a small diameter head, 9 mm chamber, with the lab machine of Figure 3.6.

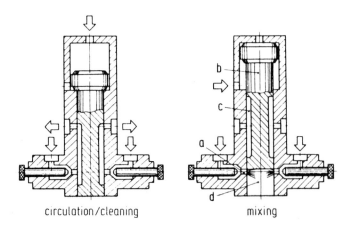

circulation/cleaning   mixing

Figure 4.1   Keuerleber and Pahl (1970) mixhead. In the closed or recirculation position. Reactants recirculate through grooves (c) along the cylindrical cleanout piston (b). In the open position. Reactants flow at high velocity through circular orifices (a), impinge in the chamber (d) and flowout to the mold cavity (diagram from Oertel, 1985).

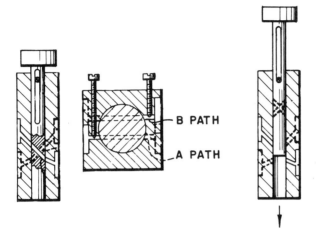

Figure 4.2   Macosko and McIntyre mixhead . Components recirculate through diagonal holes through the ram (from Macosko and MacIntyre, 1984).

Other designs employ separate valves for opening and closing the inlet nozzles. An example is shown in Figure 4.3. This design also employs slit-shaped, rather than circular orifices. The reactants recirculate across these orifices as indicated in section A of the figure. As the cleanout piston retracts, separate valves on the metering unit switch the recycle line to another feed stream. The inlet slit length can be decreased by reducing slightly the stroke of the cleanout piston.

In all designs the switching from recirculation to injection should be as fast as possible. The pressure trace shown in Chapter 3, Figure 3.4, illustrates the transients which occur as a mixhead, like the one in Figure 4.1, is opened. As the end of the mixhead ram passes over the injection nozzles, flow is stopped momentarily. Pressure builds up and the feed lines expand slightly. This expansion will prevent the flow rates from being at the proper ratio when the head is first opened. Also, as the nozzles are partially uncovered or closed again at the end of a shot, the streams may not impinge properly. Both these effects can lead to streaks and sticky regions in the product. They are often called "lead-lag" problems in mixhead operation.

In order to establish the mixing pattern in the impingement chamber, it is also necessary that the head fills quickly with liquid as flow starts. The streams may spray or spiral along the chamber walls. Some back pressure is needed to prevent this. Flow through the runner and the mold itself may be sufficient, but often a flow restriction or aftermixer is added. These are built into the runners and gate and must be trimmed from the demolded part. Figure 4.4 shows two different gate designs. The "harp" aftermixer style was used in earlier RIM molds to "catch" any poorly mixed material from the beginning of the shot. Such traps are neither very effective nor necessary with a fast acting, recirculation head. The flow diverter style is simpler to construct and also leads to less problems with trapped gas bubbles. Mikkelsen and Macosko (1988) report that a simple tapered runner with an inlet about half the mixhead diameter also works well.

An alternative to flow diverters or aftermixers built into the mold are mechanical throttling devices built into the mixing head. Figure 4.5 illustrates a simple design. The right angle bend in the flow causes enough pressure drop to fill the mixing chamber quickly. The restriction can be increased by partially blocking the chamber outflow with the second clean out piston. Such mixhead throttling devices permit smaller mixheads and simpler, smaller runner systems. They are also useful for insuring smooth flow into open molds such as those used in foaming (Molnar and Lee, 1988). Flow into molds and gate

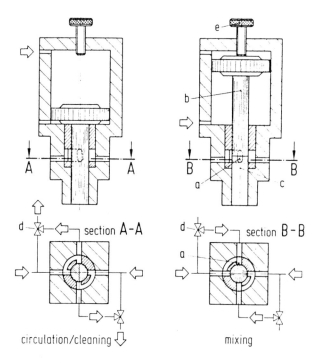

Figure 4.3   Wisbey (1977) mixhead. Recirculation of reactants is controlled by external valve (d). Injection is through four slits (a). The size of these slits is controlled by (e) which regulates the backstroke of the cleanout piston (b) (diagram from Oertel, 1985).

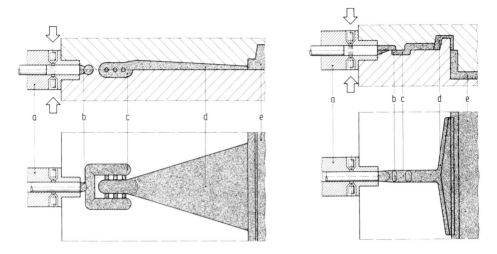

Figure 4.4   Several aftermixer designs and gating systems. Left is a "harp" aftermixer connected to a fan gate (d) which feeds the mold cavity (e). (c) is designed as a dead leg to "catch" the poorly mixed material from the initial part of the shot. Right is a flow diverter (b,c) style runner connected to a dam gate (e) (from Oertel, 1985).

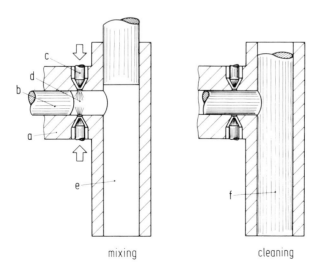

Figure 4.5   L shaped throttling system (Schülter, 1973; Florentini, 1982; diagram from Oertel, 1985).

design for closed molds will be discussed in the next chapter. A disadvantage of such throttles is that they require another mechanical motion which must be closely synchronized with the mixhead. Throttle motion can be too slow for very fast reacting formulations.

Figure 4.6 gives a relatively new and quite different mixhead design. It is a parallel stream rather than impingement mixer. Components flow through the head concentrically with the polyol in the center surrounded by an annular stream of isocyanate. The runner to the mold forms the mixing chamber so there is no clean out piston. This permits the head to open and close very quickly, perhaps an advantage for very fast reacting systems. However, its tapered sealing surfaces make it difficult to use with glass or mineral filler. As yet no data on the performance of this design or comparisons to the more common impingement types are available.

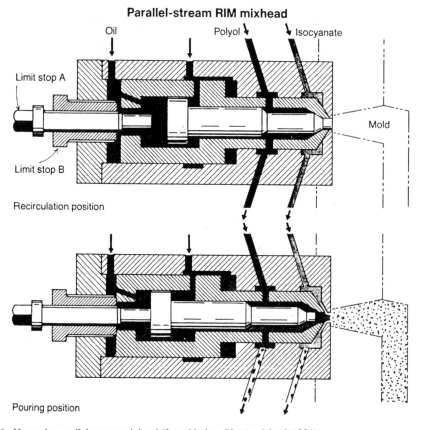

Figure 4.6   Hennecke parallel stream mixhead (from *Modern Plastics*, March 1984).

4.2 Measures of Mixing Quality

All of these mixhead designs were developed by trial and error. Even today when a new design is developed, it is put on a machine and tried with a typical formulation. Mixing quality is judged by visual appearance. If the polymer in the runner is sticky or stringy, especially when cut with a knife or if streaks, layers, or blisters appear in the part especially when it is heated in an oven, the polymer is "poorly mixed". Actually this is a sensitive but qualitative method of measuring a limit of mixedness. Some results are shown in Table 4.1. For the first entry, a crosslinking formulation F-1, raising the polyol tank temperature from 46 to 52°C improved the plaque appearance significantly. With a DETDA formulation similar to F-5, an increase in flow rate from 102 to 138 g/s produces the same type of improvement. For F-1 raising the polyol temperature lowered its viscosity making it "easier" to mix. With the F-5 the increase in flow rate put more kinetic energy into the mixing chamber.

A more quantitative way to study the influence of these variables on mixing is to express them in terms of a Reynolds number, the ratio of inertial force or kinetic energy which helps mixing to viscous force or dissipation which reduces mixing. Reynolds number can be defined for the flow of component A (or B) through its inlet nozzle in terms of its density ρ, flow rate Q, viscosity η and diameter d

$$Re_A = \frac{4\rho_A Q_A}{\pi \eta_A d_A} \tag{4.1}$$

The units of these quantities are selected so as to make Re dimensionless.

A major problem with applying Equation 4.1 to RIM mixheads is the determination of the nozzle diameter d. Most mixheads use some type of flow restrictors in or near the nozzle. The screws in Figure 4.1 are an example. Others use noncircular orifices like the slits in Figure 4.3. Thus an effective or hydraulic diameter must be described. Mrotzek (1982) and Vespoli (1984) suggest using the pressure drop through a sudden contraction to define an effective nozzle diameter.

$$\Delta p = \frac{8C_D \rho Q^2}{\pi^2 d^4} \tag{4.2}$$

If the diameter change is large, $d_0/d \gg 1$, and the flow rate is high, inertial effects dominate. Then C_D, the drag coefficient, should be constant ~ 1. The data of Figure 4.7 test

TABLE 4.1   Mixing Quality Correlated to Reynolds Number

Reference[a]	Isocyanate Stream T °C	η mPa·s	ρQ g/s	d[b] mm	Re	Polyol Stream T °C	η mPa·s	ρQ g/s	d[b] mm	Re	Appearance[c]
Manas (1988)											
F-1											
	46	(low)	100	0.5	800	46	351	150	1	240	sticky
u-MDI	46		100	0.5	800	50	290	150	1	280	streaks
PCL triol	46		100	0.5	800	52	264	150	1	300	good
	46		100	0.5	800	54	240	150	1	320	good
u-MDI	70		33	0.5	900	76	92	50	1	290	streaks
PCL triol	70		33	0.5	900	80	79	50	1	310	good
Lawler (1976)											
F-3											
u-MDI + BDO +	21	30	250	1.4	7300	~50	800	393	1.80	347	streaks, blisters
P(PO-EO)+	21	30	250	1.4	7300	~60	630	399	1.82	444	good
dispersed polym.											
Vespoli (1984)											
F-4											
p-MDI(36%)	49	302	~100		405	49	986	~120		102	layered, sticky
EG (36%)	60	205	~100		602	60	687	~120		146	runner stringy
PTMO	63	205	~100		602	66	554	~120		182	good
F-5											
p-MDI (50%)	50	123	59		676	50	350	102		290	runner stringy
DETDA(50%)	50	123	80		726	50	350	138		337	good
P(PO-EO)											
p-MDI	49	110	153	1.55	1140	49	640	358	2.37	301	sev. striations
DETDA	49	110	153	1.30	1359	49	640	358	1.99	358	v. slight
P(PO-EO)+	49	110	153	1.17	1510	49	640	358	1.80	396	good
10% milled glass	42	525	153	1.17	316	49	640	358	1.80	396	v. slight
Fruzzetti (1977)											E_{flex} (MPa)
Body Panel	20	100	366	4.2	1100	32	1400	352	2.1	150	untestable
Formulation	20	100	116	4.2	350	66	240	111	2.1	280	760
	20	100	366	4.2	1100	66	240	352	2.1	890	920-1010
											Flex life(kHz)
Shoe sole	38	250	146	4.2	180	38	470	308	4.2	200	90-140
formulation	47	170	183	4.2	650	60	190	385	4.2	615	235
	47	170	183	2.1	825	60	190	385	4.2	615	>250

a)   Formulation references, F-1 etc., indicate the closest formulation in Table 2.4

b)   All except Manas and Fruzzetti calculated based on nozzle pressure drop, Equation 4.2.

c)   All observations on ~3mm thick plaques molded with an aftermixer.

this assumption for a simple conical orifice. For the larger contraction C_D is nearly constant for Re > 100. Data for more complex nozzle designs are not available, but since d depends only on $C_D^{1/4}$, Equation 4.2 should be useful.

Applying Equation 4.1 and, where necessary, Equation 4.2 Re values can be calculated for the studies reported in Table 4.1. For the crosslinking formulation, F-1, good mixing is achieved when Re ≅ 300. For a BDO based formulation, similar to F-3 in Table 2.4, Lawler (1976) found Re ≥ 440 sufficient. His work indicates that the lowest Re, typically for the polyol, determines mixing quality. For the ethylene glycol based formulation the critical Re appears to be 180. For the DETDA system F-5 a polyol Re of about 300 is needed for good runner appearance. A similar formulation with 10% glass requires Re > 360 to eliminate visible striations. Note that here Re is increased by decreasing the effective nozzle diameter, i.e. by restricting nozzle flow somewhat by advancing a screw into the orifice (see Figure 4.1).

Final physical properties have also been used to judge mixing quality. The last two studies in Table 4.1 show such measurements by Fruzzetti, et al. (1977). For the body panel formulation flexural modulus reaches its ultimate value at a Re around 300. For the shoe sole formulation Re has an influence on flex life up to 300 to 600. This cyclic test may be particularly sensitive to a very few poorly mixed regions.

In an attempt to find a more quantitative measure of mixing and one which does not require molding a part, Richter and Macosko (1978) measured the adiabatic temperature rise (see Chapter 2.5) as a function of Re. As illustrated in Figure 4.8 for a crosslinking RIM formulation, apparent reaction rate and maximum temperature increase with increasing Re. Above a critical Re no further improvement is seen. Ranz (1979) and Chella and

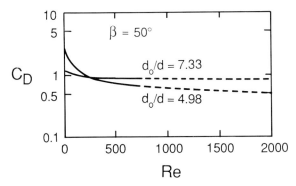

Figure 4.7   Drag coefficient vs Reynolds number for two conical nozzles, β = 50°, with different inlet to outlet diameters (Mrotzek, 1982; adapted from Müller et al. 1984).

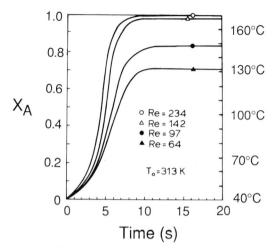

Figure 4.8   Adiabatic temperature rise for a crosslinking polyurethane RIM formulation F-1. Re increased by increasing flow rate (adapted from Chella and Ottino, 1983).

Ottino (1983) explain these curves qualitatively in terms of a lamellar model. The mixing flow generates lamella with a distribution of striation thicknesses. Where the striations are thin, monomers diffuse together and form polymer. But if some of the striations are large, they will become trapped between the polymer layers. This will cause local stoichiometric imbalance and thus a slower rate and lower maximum temperature rise. Increasing Re re-

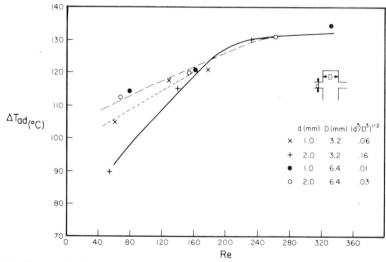

Figure 4.9   Maximum adiabatic temperature rise vs Re for several mixhead geometries, similar formulation to Figure 4.8 (from Lee, et al., 1980).

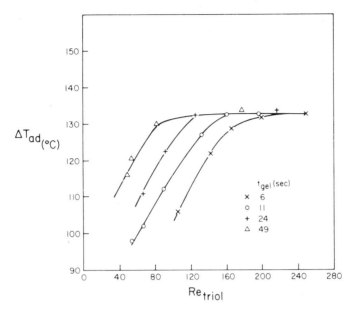

Figure 4.10   Maximum adiabatic temperature rise vs Re for four different catalyst levels (from Lee, et al., 1980).

duces the number of "too large" striations until eventually all the monomer can react and homogeneous kinetic data are obtained. Chella and Ottino's lamellar diffusion and reaction model indicates that for reaction speeds on the order of 10 s striations must be less than 10 μm for the temperature rise to be controlled by reaction kinetics.

Lee, et al. (1980) used the adiabatic method to evaluate some different mixhead geometries. Figure 4.9 shows that Re_c did not change with nozzle and chamber dimensions. Similar results indicated no difference between 180° and 120° impingement angle. Re_c was found to increase with catalyst level, Figure 4.10. This can be explained qualitatively: the faster reacting systems allow less time for diffusion and thus need a higher Re to produce thinner striations.

Figure 4.11 shows the same type of measurements for a phase separating BDO formulation, F-2 (Kolodziej, et al., 1982). The adiabatic temperature rise appears much less sensitive for this formulation than for the crosslinking one in Figure 4.8. However, the molecular weight distribution of these samples shows large differences, Figure 4.12. As Re increases molecular weight increases and the distribution narrows. This would be expected for a polymerization which goes from heterogeneous to a more homogeneous one.

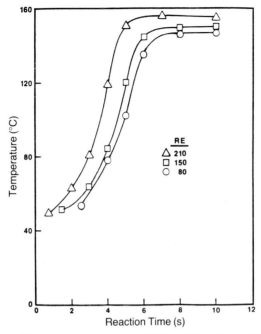

Figure 4.11   Adiabatic temperature rise for a BDO RIM formulation, F-2 (from Kolodziej, et al., 1982).

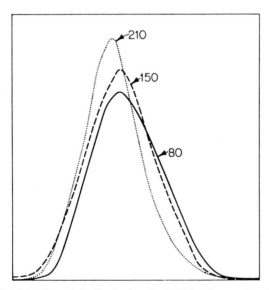

Figure 4.12   Molecular weight distributions by GPC for the samples of Figure 4.11 (from Kolodziej, et al., 1982).

The adiabatic temperature rise is less sensitive because 80% of the temperature rise in this formulation is due to BDO, a small molecule which can diffuse readily to reactive sites. Molecular weight on the other hand is dominated by incorporation of the long oligomer into the chain.

Molecular weight analysis methods like GPC will not work on most commercial RIM formulations because they use trifunctional oligomers and thus are crosslinked. Measurement of swelling and fraction of extractables can be used to quantify mixing quality. Higher swelling or more extractables means less complete reaction and thus poorer mixing (Willkomm, 1989).

Another method which has been used to quantify mixedness is emulsification. Begeman and coworkers (1986) impingement mixed polyol with glycerine. They centrifuged the resulting emulsion at about 3000 rpm and then recorded the percent of the sample volume which was still emulsified. Figure 4.13 shows some of their results. It appears that maximum emulsification is achieved at Re about 300.

A well known method for measuring mixing quality is to extract a number of small samples from the mixture, measure the concentration of a tracer or reactant in each sample, and compute the variance of these concentrations. The smaller the variance, the better the mixture. Malguarnera and Suh (1977), Tucker and Suh (1980) and Mrotzek (1982) have used this method to evaluate impingement mixing. Figure 4.14 illustrates typical results. All three studies indicate strong reduction in variance up to Re = 300 and little improvement above 500. Tucker and Suh point out that the size scale resolution of this technique is $\geq$ 200 μm.

4.3   Visualization

The methods described above indicate that impingement mixing improves up to Re = 200 to 500. However, they are all indirect; they don't give us information about what is occurring inside the mixing chamber. Some direct studies of fluid mechanical mixing in the impingement process are available.

Several workers have measured flow patterns in the impingement chamber. Tucker and Suh (1980) injected dyed water and glycerine/water mixtures into a transparent chamber. Some of their results are reproduced in Figure 4.15. Below Re $\cong$ 60 each stream hugs its side of the chamber in laminar flow. Above Re 60 the flow becomes unstable. Flow becomes so chaotic by Re $\cong$ 150 that the chamber appears black. Lee, et al. (1980) also observed departure from laminar flow at Re $\cong$ 50 and chaotic flow near 150.

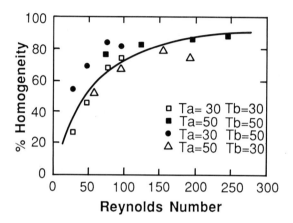

Figure 4.13   Emulsification of polyol (a) glycerine (b) mixture vs Re at several temperatures and flow rates; pressure held constant (replotted from Begemann et al., 1986, assuming effective d = 1 mm).

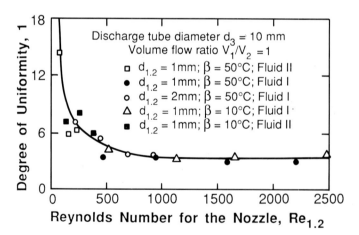

Figure 4.14   Uniformity of a mixture of a clear and a dyed polyol vs Re for several conical nozzles and two viscosity levels (replotted from Müller et al., 1984).

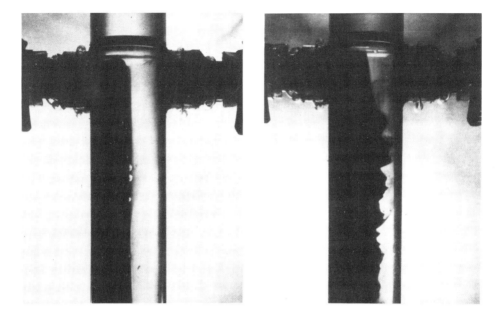

Figure 4.15   Low Re visualization with one stream dyed. a) Laminar flow persists up to about Re ≅ 60.
b) Above 60 flow becomes unstable, eddies form (from Tucker and Suh, 1980).

To better visualize the high Re regime Sandell, Macosko and Ranz (1985) injected a fading dye into the center of each stream. Some of their results are shown in Figure 4.16. The dye is thymol blue which when well mixed with the slightly alkaline water becomes clear. This is an acid-base reaction and essentially instantaneous. We see a large change between Re = 230 and 515. At 155 and 230 there are dark streaks all the way to the bottom of the photographs. In the mirror on the left of each photo we can see that the flow is not well distributed between front and rear of the chamber. At Re = 515 the color distribution is much more uniform and is nearly gone at the bottom of the photo. This suggests nearly complete mixing and diffusion of acid and base within one chamber diameter of the impingement point. Sandell's results showed little improvement in mixing when Re was increased from 400 to 700. All the visualization studies indicate that the turbulence produced by impingement reduces to laminar flow (and thus much less efficient mixing) within two or three diameters of the impingement point.

Bourne and coworkers (1981) have developed an azo coupling reaction which generates two different dyes in proportions dependent on the mixing history. Shannon and coworkers (1986) have used this reaction to evaluate impingement mixing quality and also find little improvement for Re > 300.

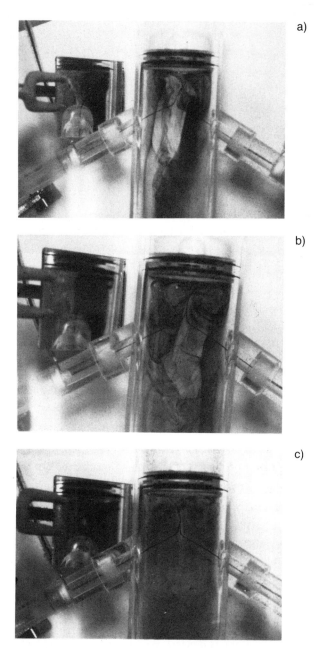

Figure 4.16 High Re visualization with injection of a fading dye in the center of each stream. In the left of each photograph is a mirror to show back to front dispersion of the tracer (from Sandell et al., 1985).

Kolodziej, Macosko and Ranz (1982) used polymerization to freeze the mixing patterns. They dispersed 0.1% carbon black in the polyol of a BDO formulation (F-2, Table 2.4) which was then injected vertically into a tapered tubular mold. After polymerization the samples were collected from the center of the tube, sectioned and photographed, Figure 4.17. Near the center there will be little laminar shearing after leaving the impingement mixing zone; thus these micrographs should be a good measure of the striations created by impingement. Random lines were drawn across the micrographs and the distances between two black regions measured. These are plotted in Figure 4.18.

Baldyga and Bourne (1983) have modelled the data shown in Figure 4.18 theoretically. Assuming that impingement mixing produces isotropic, homogeneous turbulence in the region around the impingement point (two chamber diameters) and that the residence time distribution of fluid flowing through this turbulent zone is exponential, they obtain the curve in Figure 4.18. The fit to the experimental results of Kolodziej is quite good; there are no adjustable constants. Bourne and Garcia-Rosas (1985) point out that flow through the runner and into the mold will continue to reduce striations. However, because this flow is laminar it will not reduce striations in the center.

The results in Figure 4.18 show that mixing improves (i.e. striation thickness decreases) with Re only up to about 250. This is in qualitative agreement with all the other studies; there is little influence of Re above a value 200 - 500. Another significant conclusion from Figure 4.18 is that the average striation thickness is about 100 μm. This is of the same order as found by Tucker and Suh (1980) in their color variation work. Nguyen and Suh (1985) went to Re = 4000 to make interpenetrating network polymers. Even at this high Reynolds number their micrographs indicate a size scale of 30μm. These values are much larger than the 1μm needed for molecular diffusion to be able to wipe out concentration gradients in the time scale of typical RIM reactions. Another mixing mechanism besides impingement must be responsible for the high polymers produced by RIM.

4.4   Micromixing

The results of the visualization studies described above show that the mixing created by impingement is itself insufficient for the fast polymerizations in RIM. Yet high molecular weight polymers are produced; RIM does work. Some other mechanism must be responsible for reducing component separation from 100μm to the ~1μm needed for kinetic control.

Ranz (1979) suggested that the molecular interface between two monomers might become unstable as they react due to density changes or due to the reaction products acting

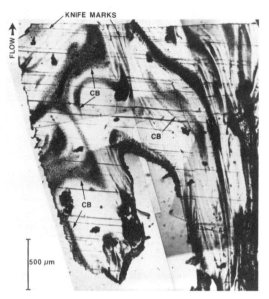

Figure 4.17 A microtomed sample from the center of a tube molded with carbon black doped polyol (from Kolodziej, et al., 1982).

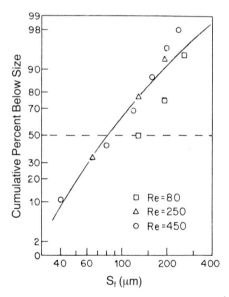

Figure 4.18 Cumulative size distribution of striations at various Re, (Re = 80), Δ (250) O (450) from photos like Figure 4.17 (Kolodziej, et al., 1982). Line from isotropic turbulence model of Baldyga and Bourne (1983).

as a surfactant as with spontaneous emulsification. Fields, Thomas and Ottino (1986) observed surface waves on a 100µm level in the interface of u-MDI with BDO and with PPO. Wickert, Macosko and Ranz (1987) looked at similar systems at higher magnification and found evidence for smaller scale dispersion. Machuga et al. (1988) used a video microscope to study the interface between a number of polyols and polyamines with isocyanates. They observed very strong interfacial activity. Figure 4.19 shows a sequence of frames from the video microscope during the first 2.5s after contact between an amine terminated polyether and MDI. The interfacial region grows to over 100µm in 0.8s. After 20s this region was isolated and found to have $M_w \approx 16,400$. Thus this interfacial activity is able to wipe out the striations left after impingement and build high polymer on the time scale of the RIM process.

Higher magnification studies suggest that interface consists of many tubules of the diamine shooting into the MDI. Machuga et al. propose that these result from local imbalance in internal pressure at the interface. A pressure difference of < 0.1 bar is sufficient to give an initial growth rate of 20-30 µm in the first 0.1s. The pressure difference at the tip of each tubule keeps driving more diamine into the MDI until the polymer growing on the tube wall behind it chokes off the flow.

All systems that they studied seemed to have about the same initial interfacial growth rate. This included the polyols and polyamines with isocyanates, diamine with diacid chloride and even nonreactive but dissimilar systems. However, there was a great difference in the ultimate thickness of the interfacially mixed phase. Combinations which react rapidly to form highly crosslinked polymers or polymers which precipitate seem to choke off the tubules more quickly. These observations correlated with systems for which it is hard to produce a polymer via RIM. Figure 4.20 compares interfacial thickness measured from video frames vs time for an aliphatic triamine contacted with two different diisocyanates. Both grow to over 50 µm in the first 0.5s but the faster reacting aromatic diisocyanate quickly slows down and stops at about 90 µm while the slower reacting aliphatic diisocyanate continues to grow.

In summary, the results of research into impingement mixing for RIM suggest the following model. The fluid mechanics of impingement reduces the scale of segregation from the diameter of the inlet nozzles (1-3mm) to a scale of order 100µm. This reduction does not depend significantly on the design of the mixhead. Segregation decreases from 1mm to 100µm with Re up to a critical value, around 300. Further increase in Re has little effect.

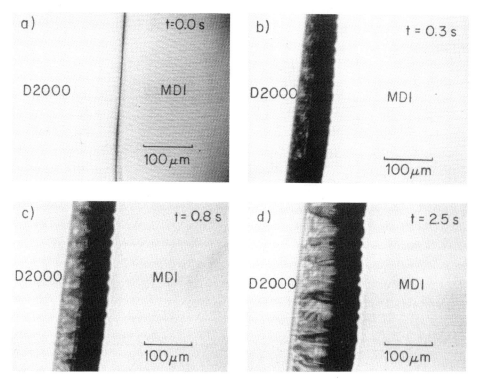

Figure 4.19 Spontaneous growth of an interfacial mixed phase upon contact of amine terminated polypropylene oxide (D2000, Texaco) and a uretonemine modified MDI (LF 168, ICI). Frames are from a video microscope at 25°C. a) At t = 0s, point of contact. Reactants were brought together slowly between glass slides 250 μm apart (from Machuga et al., 1988).

Further reduction in scale results from fluid mechanical mixing during flow in runner and mold but particularly from physiochemical effects at the interfaces between the two reactant streams. A pressure at the interface seems to drive one reactant rapidly into the other through tubules. The length of these tubules and thus mixing is governed by reaction speed and properties of the resulting polymer. Finally, molecular diffusion takes over to bring all the components into molecular contact and complete the reaction. These three mechanisms are illustrated schematically in Figure 4.21.

Since the fluid mechanical means to achieve critical Re are readily available for many reactive systems of interest, reaction *micromixing* is the mechanism which we need to control. More research is needed to determine the role of each of the variables mentioned above in different space and time scales.

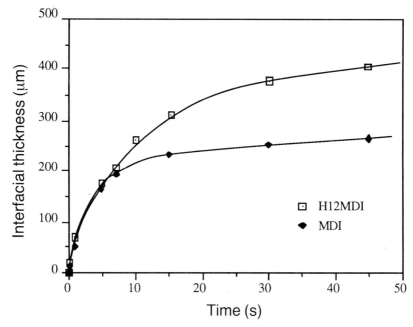

Figure 4.20 Thickness of the interfacial mixed phase vs. time for a mixture of aliphatic triamines (T403: T5000, 1:1 by weight, Texaco) with u-MDI and methylene bis(4-cyclohexyl isocyanate) H$_{12}$MDI (from Machuga, 1988).

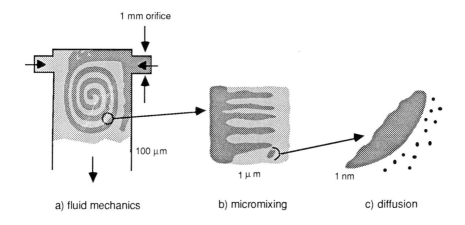

Figure 4.21 The three mechanisms active in impingement mixing.

REFERENCES

Baldyga, J; Bourne, J.R. "Distribution of striation thickness from impingement mixers in reaction injection molding," *Polym. Eng. Sci.*, **1983**, *23*, 556-559.

Begemann, M.; Maier, U.; Müller, H.; Pierkes, L. "RIM-improved technology of the machines and selective mould design enhance moulding quality," from proceedings of 13th Plastics Colloq., Technischen Hochschule Aachen, 1986.

Bourne, J.R.; Garcia-Rosas, J. "Laminar shear mixing in reaction injection molding," *Polym. Eng. Sci.* **1985**, *25*, 1-5.

Bourne, J.R.; Kozicki, F.; Rys, P. "Mixing and fast chemical reactions to determine segregation," *Chem. Eng. Sci.*, **1981**, *36*, 1643-1648.

Chella, R.; Ottino, J. M. "Modelling of rapidly-mixed, fast-crosslinking, exothermic polymerizations," *AIChE. J.* **1983**, *29*, 373-382.

Chen, C.H.Y.; Briber, R.M.; Thomas, E.L.; Xu, M.; MacKnight, W.J. "Structure and morphology of segmented polyurethanes: 2. Influence of reactant incompatibility," *Polymer* **1983**, *24*, 1333-1340.

Fields, S.D.; Thomas; E.L.; Ottino, J.M. "Visualization of interfacial urethane polymerization by means of a new microstage reactor," *Polymer* **1987**, *27*, 1432.

Florentini, C.; "Head for mixing and ejecting interacting liquid components for molding plastic articles," US Patent 4,332,335 (1982), Afros-Canon.

Fruzzetti, R.E.; Hogan, J.M.; Murray, F.J.; White, J.R. "Factors affecting the quality of impingement mixed RIM urethanes" Soc. Auto. Eng. meeting, Detroit, Paper No. 770839, Sept. 1977.

Kolodziej, P.; Macosko, C.W. ; Ranz, W.E. "The influence of impingement mixing on striation thickness distribution and properties in fast polyurethane polymerization," *Polym. Eng. Sci.* **1982**, *22*, 388-392.

Kuerleber, R.; Pahl, F.W."Device for feeding flowable material to a mold cavity," German Patent 2,007,935 (1970); U.S. Patent 3,706,518 (1972) Kraus Maffei.

Lawler, L.F. Union Carbide Corporation, S. Charleston, W. Va, personal communication, 1976.

Lee, L.J.; Ottino, J.M. ; Ranz, W.E.; Macosko, C.W. "Impingement mixing in Reaction Injection Molding," *Polym. Eng. Sci.*, **1980**, *20*, 868-874.

Machuga, S.C. "Interfacial mixing in fast polymerizations," MS Thesis, University of Minnesota, 1988.

Machuga, S.C.; Midje, H.L.; Peanasky, J.S.; Macosko, C.W.; Ranz, W.E. "Microdispersive interfacial mixing in fast polymerizations," *AIChE J.* , **1988**, *34*, 000.

Macosko, C.W.; McIntyre, D.B.; "RIM mixhead with high pressure recycle," US patent 4,473,531 (1984).

Malguarnera, S.C.; Suh, N.P. "Liquid injection molding I. An investigation of impingement mixing,"*Polym. Eng. Sci.* **1977**, *17*, 111-115.

Manas-Zloczower, I.; Macosko, C.W. "Moldability diagrams for reaction injection molding of a polyurethane crosslinking system," *Polym. Eng. Sci.* **1988**, *28*, 000.

Molnar, J.A; Lee, L.J. "Mixing study of L-Shape mixhead in RIM process," presented at Annual Meeting of the Composites Institute of the Society of the Plastics Industry, Cincinnati, February, 1988.

Mikkelsen, K.J.; Macosko, C.W. "Characterization of laboratory RIM machines," *J. Cellular Plastics*, **1988**, *24*, 000.

Mrotzek, W.; Doktor-Ing. Thesis, Fakultät Maschinenwesen, Rheinisch-Westfallischen Technischen Hochschule Aachen, 1982.

Müller, H.; Küper, B.; Maier, U.; Pierkes, L. "The latest in molding polyurethane," *Ad. Polymer Tech.* **1984**, *5*, 257-304.

Nguyen, L.T.; Suh, N.P. "Effect of high Reynolds number on the degree of mixing in RIM processing," *Polym. Proc. Eng.* **1985**, *3*, 37-56.

Oertel, G., Ed. "Polyurethane Handbook," Hanser: Munich, 1985, Chapter 4.

Pahl F.W.; Schlüter, K. "Processing fundamentals for molded foam polyurethanes I. Storage, metering and mixing of the components," *Kunstoffe*, **1971**, *61*, 540-544.

Ranz, W.E. "Applications of a stretch model to mixing, diffusion and reaction in laminar and turbulent flows," *A.I.Ch.E. Journ.*, **1979**, *25*, 41.

Richter, E.B.; Macosko, C.W. "Kinetics of fast (RIM) urethane polymerization," Polym. Eng. Sci., **1978** *18*, 1012-1018.

Sandell, D.; Macosko, C.W.; Ranz, W.E. "A new method for impingement mixing at representative Reynolds numbers," *Polym.Process Eng.*, **1985** *3*, 57-70.

Schliter, K.; "System for filling a mold with reactive synthetic-resin components," US patent 3,975,128 **1976**, Kraus Maffei

Shannon, D.; Haese, N.; Lang, D.; Knoechel, D. "Application of the Bourne reaction-dye micromixing method to study mixing in RIM," presented at AIChE Annual Meeting, Miami, November 1986.

Tucker, C.L.; Suh, N.P. "Mixing for reaction injection molding. I. Impingement mixing of liquids," *Polym. Eng. Sci.*, **1980** 20, 875-886.

Vespoli, N.P. Upjohn Co., New Haven, CT, personal communication 1984.

Wickert, P.D.; Macosko, C.W.; Ranz, W.E. "Small scale mixing phenomena during reaction injection moulding," *Polymer*, **1987**, *28*, 1105-1110.

Willkomm, W.R.; Ph.D. Thesis, University of Minnesota, **1989**.

Wingard, R.D.; Leidal, S.M., U.S. Patent 4,085,512 (1978); also German Patent 2,607,641 (1976), Admiral.

Wisbey, J.D., U.S. Patent 4,043,486 (1977), Cincinnati Milacron.

5

MOLD FILLING

In the previous chapter we examined the mixing step in RIM. For mixing low viscosity is an advantage. Lower flow rates and thus lower pressure machines are needed to achieve the same critical Reynold's number for good mixing. For molding low viscosity is also an advantage. It means that low pressures can be used to fill large molds. Molds may be constructed from light weight materials often at a lower cost than for thermoplastic injection molding (TIM). Mold clamping forces are much lower, requiring lower cost presses.

But low viscosity also presents problems. If flow into the mold cavity is too rapid, air may be entrained and large bubbles appear in the final part. Bubbles are perhaps the greatest source of scrap production in RIM. Low viscosity reactants will wet and penetrate mold surfaces which can lead to difficult mold release. In TIM high pressure is used to pack the mold and compensate for shrinkage as the part cools. This is impossible in RIM because the low viscosity reactants will leak out the mold vents. As will be discussed in Chapter 6, foaming is usually used to balance shrinkage.

Thus low viscosity is both a key advantage for RIM and a disadvantage. To prevent air entrainment molds should be filled slowly, but for fast reacting systems this can lead to a short shot. In this chapter we will investigate how process parameters and the reaction rate should be controlled to have smooth, laminar flow yet fill the mold before the reactants gel. To do this we first examine how molds fill and then calculate how the reaction proceeds during filling.

5.1 Laminar Filling

Figure 5.1 shows the top view of a series of partial shots taken from a simple rectangular mold (Broyer et al., 1978). Flow is first radial from the gate and then becomes rectilinear as the flow front touches the cavity side walls. Since RIM machines operate at constant velocity the average flow front, L/L_0, moves linearly with the filling time.

Let us take a closer look at the flow front, now from the side of the cavity rather than the top. The flow can be divided into two regions: a parabolic velocity profile through most of the mold and fountain flow very close to the front. This is well illustrated in the recent work of Coyle et al. (1987) shown in Figure 5.2. As the flow starts the tracer assumes a parabolic profile, but since the velocity at the center of a rectangular cavity is 1.5

times that of the front (which moves at the average velocity) the tracer rapidly overtakes the front. Thus it must move away from the center, spilling out like a fountain to the walls. This flow is important for RIM because it determines the location of each reactive fluid element during filling.

The tracer positions of Figure 5.2 can be accurately modeled by finite element calculations as illustrated in Figure 5.3. However, such sophisticated calculations consume considerable computer time. Thus complete models for mold filling, which include heat transfer and reaction, have used various approximate methods to treat the flow front. All assume a flat front (90° contact angle). Castro and Macosko (1982) wrote a series expansion for the two dimensional front flow which agreed well with the finite element analysis. Domine and Gogos (1982) and Manzione (1981) used simple geometric rearrangement schemes to handle the flow front in their marker and cell codes. Lekakou and Richardson (1986) used an implicit finite difference method with a moving mesh. They solved the fully two dimensional front flow for three mold geometries: cylinder, rectangle and disk. Manas-Zloczower, et al. (1987) approximated the fountain flow as a line at the flow front which relocates fluid elements according to mass conservation. Figure 5.4 shows results of this calculation for the residence time distribution during filling. Note that the oldest material, 0.9 to 1 t_{fill}, lies in a band near but not at the wall. These compare well to values from the finite element solution except very close to the front. This simple analysis is easy to incorporate into a complete filling model and should be adequate to determine conversion distribution.

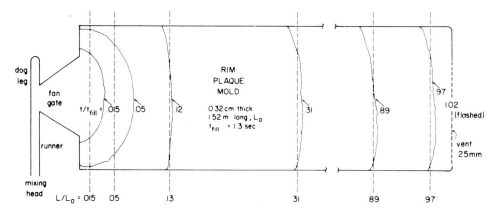

Figure 5.1   Filling of a plaque mold, top view. Figure made by tracing the front position from a series of shots made at less than the fill time, 1.3s. Mold dimensions: 1520 x 300 x 3.2mm. Filled at 1.25m/s with 900 MPa·s initial viscosity (from Broyer et al., 1978).

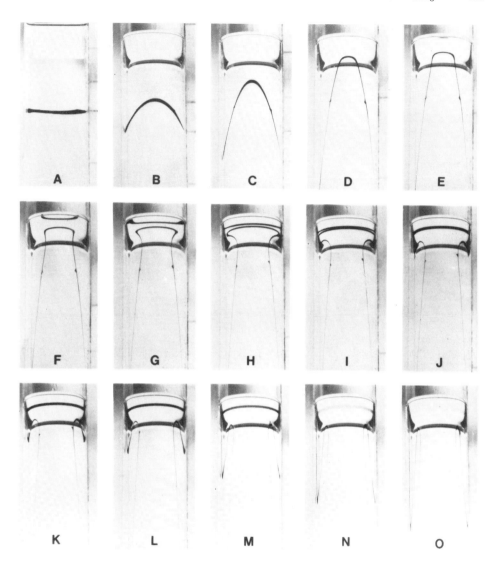

Figure 5.2   Filling of a vertical, rectangular cavity. A series of side views showing motion of a tracer injected close to the flow front. Flow is from bottom to top of each picture. Camera is moving
with flow. Because photographs are taken at an angle from below we see both menisci of the
flow front and a later reflection of the tracer. Mold thickness: 38 mm; average velocity: 1.9
mm/s; silicone fluid with viscosity 60 Pa·s (from Coyle, Blake and Macosko, 1987).

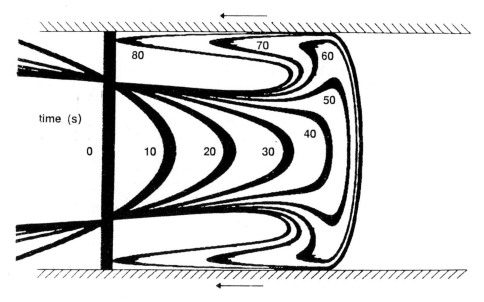

Figure 5.3   Calculated tracer position for the mold filling experiments shown in Figure 5.2 using the Galerkin finite element method (from Coyle, Blake and Macosko, 1987).

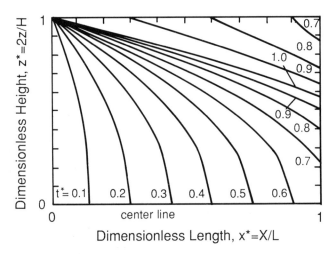

Figure 5.4   Lines of constant residence time distribution at the end of mold filling using the approximate analysis of Manas-Zloczower, Blake and Macosko (1986).

5.2 Unstable Filling, Air Entrainment

The fountain flow shown in Figure 5.2 is stable, but at high velocities it is known that the free surface of the front no longer moves smoothly into the mold cavity. This is illustrated in photos taken from the films of Castro et al. (1980), Figure 5.5. As the filling velocity is increased from 0.4 to 0.8 m/s the flow front becomes unstable. Waves appear and fingers occasionally shoot out from the front.

One mechanism for flow front instabilities is surface tension effects. We can evaluate this possibility with the capillary number, the ratio of viscous to surface tension forces:

$$Ca = v \eta / s \qquad\qquad (5.1)$$

For a typical surface tension, s=20 dyne/cm, the experiments of Figure 5.5 (velocity, v=0.5m/s, and viscosity, η=600mPa·s) will have Ca≈15. This indicates that viscous forces are dominant and that inertial effects, as reflected by the Reynolds number and perhaps viscosity gradients across the mold, are more likely the cause of these flow front instabilities.

The Reynolds number for filling a rectangular cavity is

$$Re = \frac{\rho v H}{\eta} \qquad\qquad (5.2)$$

Using mold thickness, H=3.2mm, and typical viscosity and density gives Re≈3 for the transition to unstable flow in Figure 5.5.

The polyurethane plaques produced from the experiments shown in Figure 5.5 did have some large air bubbles. Large bubbles particularly seem to occur when reactants flow around inserts or from the gate into the mold cavity. Lawler (1980) suggested that the low pressure region caused by flow around an insert can lead to coalescence of the small bubbles present in the reactants. He observed regularly spaced, large bubbles in the wake of cylindrical inserts at Re≈3.

Other studies also indicate Re criteria for air entrainment. For some time it has been recognized by RIM mold designers that there is a maximum filling rate. Knipp (1973) showed that jetting can occur as material flows from the gate into the mold. Becker (1979) recommends maximum velocities leaving film gates (Figure 4.4): v ≤ 2 m/s for 6-10mm thick cavities and 5m/s for 3-4mm. With a typical viscosity of 500MPa·s and density of 1g/cc both points translate to Re≈30. Within the gate itself Mobay's RIM structural foam manual (1978) recommends v≤1.5m/s at 1-2mm thickness. This leads to

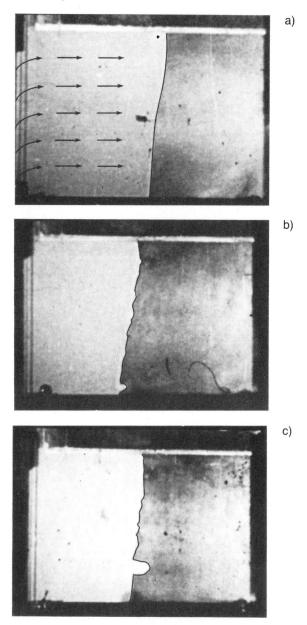

Figure 5.5 Filling, top view, taken through a polycarbonate top mold surface. As average velocity is increased from a) 0.4 to b) 0.5 and c) 0.8m/s, fingers begin to shoot out from the flow front. Mold dimensions: 457 x 305 x 3.2mm; RIM 2200 (F-3, Table 2.4) with initial viscosity 600MPa·s (from 8 mm movie film by Castro et al., 1980, flow front position traced over with pen).

Re≤5. For flow out from direct fill, radial gates (Figure 5.21) they recommend Re≤100. Clearly specific gate design can have an influence and will be discussed later in this chapter.

5.3 Analysis of Mold Filling

The results discussed above indicate that smooth, laminar filling can be achieved by operating below some critical Re which is in the range 10-100. We now need to determine whether the mold can be filled under these restrictions, i.e. how much reaction occurs during filling.

Obviously if the time needed for filling is much less than the reaction time, then no reaction will occur during filling. This is illustrated in Figure 5.6. Here the filling time is typical for high production RIM, about 1s, while the adiabatic gel time is much longer, 5s. The thin mold, 1.6mm, requires much more pressure to fill than the thick one, 3.2mm, but even these are very low, <2bar. For the thin mold there is some deviation from the linear pressure rise expected during isothermal filling. This is due to heat transfer to the hotter mold walls.

Figure 5.7 shows pressure rise during filling of a thicker mold with a faster reacting system. As the filling time is increased from 1.2 to 3.2s we see a dramatic increase in pressure. This is due to high conversion during filling which causes the viscosity to increase. This viscosity rise is quite rapid and quickly leads to gelation which prevents the mold from being completely filled as shown in Figure 5.8.

Pressure during mold filling can be predicted using the momentum balance. For steady, laminar flow into a wide rectangular cavity this balance reduces to (Castro and Macosko, 1982)

$$\frac{\partial p}{\partial x} = \frac{\partial}{\partial z} \left(\eta \frac{\partial v_x}{\partial z} \right) \tag{5.3}$$

This result assumes unidirectional flow down the mold cavity, x direction, and ignores any density change (foaming) that might occur during filling. Using the overall mass balance velocity can be replaced with the constant flow rate, Q. Integration gives

$$p = \frac{4Q}{W} \int_0^{x_f} \frac{dx}{\int_0^H \frac{z^2 dz}{\eta}} \tag{5.4}$$

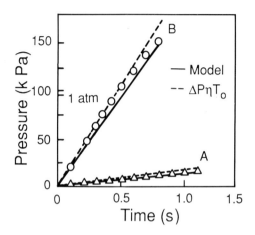

Figure 5.6   Pressure vs filling time at the entrance to a rectangular mold cavity.   Prototype BDO polyurethane (F-2 in Table 2.4) with a lower catalyst level which gives an adiabatic gel time of about
5s at the initial temperatures used.   Points represent experimental data, dashed line constant
viscosity during filling, solid line includes changes in viscosity due to heat transfer and reaction.   Mold A: 425 x 101 x 3.2mm, Q=124cc/s, T_o=49.3°C, T_w=74.1°C; mold B: 710 x 101
x 1.6mm, Q=145 cc/s, T_o=52.5°C, T_w=65.2°C (adapted from Castro and Macosko, 1982).

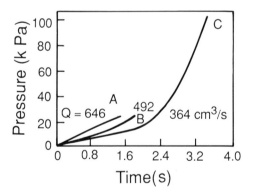

Figure 5.7   Experimental pressure vs filling time for RIM 2200, adiabatic gel time about 2s.   Mold: 457 x
305 x 6.4mm, T_w=65 °C; curve A: Q=646cc/s, T_o=52°C; curve B: 492cc/s, 51°C;   curve C:
364cc/s, 60°C (adapted from Castro et al., 1980).

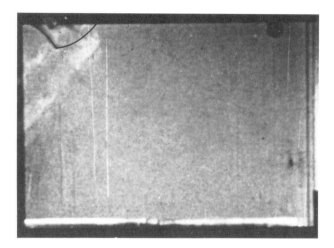

Figure 5.8   Photo of filling with premature gelation leading to a "short shot." Same conditions as curve C in Figure 5.7.

If viscosity is constant, i.e. no reaction or heat transfer during filling, this equation is easily integrated to give the familar result for Poiseuille flow in a rectangular channel

$$p = 12\eta \, Qx_f / WH^3 \qquad (5.5)$$

Pressure is linearly dependent on the flow front position, x_f, and thus the filling time. Equation 5.5 predicts well the data in Figure 5.6.

If, however, there is significant reaction or heat transfer during filling, then viscosity is not constant but a function of temperature and conversion at each location in the mold. Viscosity functions are given in Chapter 2 for crosslinking and for phase separating polymerizations. For example, for several phase separating urethane systems Castro et al. (1984a) have correlated viscosity with the temperature dependence of the initial viscosity, η_o, and conversion of isocyanate groups, α.

$$\eta = \eta_o(T) \left(\frac{\alpha_g}{\alpha_g - \alpha} \right)^{C_1 + C_2 \alpha} \qquad (2.43)$$

where $\eta_o(T) = A_\eta \exp(E_\eta / RT)$.

To determine temperature and conversion we need to use the energy and mole balances. Neglecting diffusion and viscous dissipation, assuming a second order reaction, constant thermal properties and heat conduction only across the mold thickness, z, these balances become (Castro and Macosko, 1982; Manas-Zloczower and Macosko, 1988)

$$\frac{\partial T}{\partial t^*} + v_x \frac{\partial T}{\partial x^*} = \frac{1}{Gz} \frac{\partial^2 T}{\partial z^{*2}} + \frac{Da}{Gz} A^* \exp(-E_a/RT) (1 - \alpha)^2 \qquad (5.6)$$

    convection   conduction   reaction

$$\frac{\partial \alpha}{\partial t^*} + v^*_x \frac{\partial \alpha}{\partial x^*} = \frac{Da}{Gz} A^* \exp(-E_a/RT) (1 - \alpha)^2 \qquad (5.7)$$

    convection   reaction

where Da = Damköhler number = $\dfrac{\text{heat produced by reaction}}{\text{heat transfer by conduction}}$

$$= \frac{C_o k (T_o)}{4a/H^2}$$

Gz = Graetz number = $\dfrac{\text{heat transfer by convection}}{\text{heat transfer by conduction}}$

$$= \frac{(v_x/L = 1/t_f)}{4a/H^2}$$

and   a = thermal diffusivity = $k/\rho C_p$

These equations have been made dimensionless in the usual way: $x^* = x/L$; $z^* = 2z/H$; $v^* = v_x/\overline{v_x}$; $t^* = t \cdot 4a/H^2$. If $A^* = \Delta T_{ad} A/k (T_o) = \Delta T_{ad} [\exp (-E/RT_o)]^{-1}$ then it is not necessary to make temperature dimensionless. There are other dimensionless forms for these equations; see for example Chapter 6.

Equations 5.6 and 5.7 are coupled to the momentum balance 5.4 through the viscosity 2.43. Thus all four must be solved simultaneously. This has been done by Castro and Macosko (1982) using finite difference methods. Some of their results for pressure rise during mold filling are compared to experimental data in Figures 5.6, 5.9 and 5.10. In 5.6 heat transfer from the mold wall for the thin mold leads to a slightly decreasing pressure rise. The same trend is predicted by the nonisothermal filling model. In 5.9 and 5.10 filling is much slower and there is time for significant reaction to occur. The model predicts the rapid build up of viscosity and thus pressure. Note that if an approximation for the fountain flow is not included in the model, the pressure rise is seriously under-predicted. Figure 5.11 shows the calculated conversion and temperature profiles across the mold at several positions at the end of filling. Note that the maximum in conversion corresponds to the line of maximum residence time, Figure 5.4, except near the inlet, $x^*=0.05$. This deviation is due to heat conduction from the wall. The temperature calculations compare well to thermocouple measurements as shown in Figure 5.12.

5.4 Adiabatic Filling

Because RIM molds are filled rapidly and polymers are poor heat conductors, it is often possible to neglect heat transfer during filling. The fact that maximum conversion and maximum residence time are similar in Figure 5.11 is one evidence for this assumption. Another is shown in the data of Vespoli and Alberino (1985) who measured pressure during filling at the entrance to a long runner feeding a rectangular cavity, Figure 5.13. They found essentially no difference in the pressure rise when the mold wall temperature was increased from 50 to 77°C and even at 154° the slope increased only slightly. Yet in all three cases there was considerable reaction during filling. Figure 5.9 and 10 also show similar pressure rises despite a 12°C difference in mold temperature.

The case for adiabatic filling is very strongly supported by the experiments of Vespoli et al. (1986) on a polyurea formulation. Steady state temperature measurements during mold filling, Figure 5.14, agree very well with adiabatic temperature rise for the same reactants after a time equal to the average residence time, Figure 5.15. There were, however, no pressure data at different mold temperatures as in Figure 5.11.

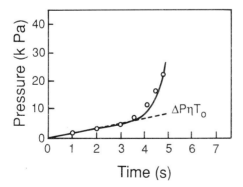

Figure 5.9 Pressure vs filling time for prototype BDO. Mold: 505 x 101 x 3.2mm, $T_w = 82$°C, Q = 33.5cc/s, $T_o = 55$°C (adapted from Castro and Macosko, 1982).

The above results suggest a simple approach to determine whether or not a given formulation can fill a given mold. If the time to fill the mold exceeds the adiabatic gel time, flow will stop (Manas-Zloczower and Macosko, 1988). Thus

$$t_f < t_{g,ad} \tag{5.8}$$

is a criterion for RIM mold filling. We expect this to be a conservative criterion since according to Figure 5.4 only an infinitesimally thin line will have gelled at $t_f = t_{g,ad}$. This point is probably closer to the beginning of the strong pressure rise.

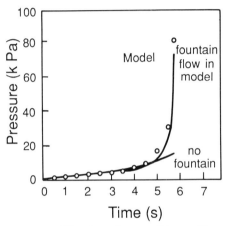

Figure 5.10   Pressure vs filling time. Same conditions as Figure 5.9 except Q=27.5cc/s and $T_w = 70°C$.

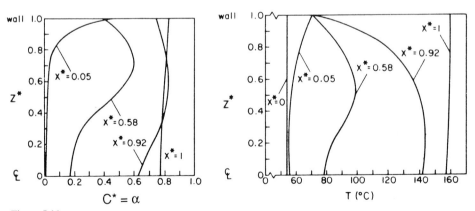

Figure 5.11

a)   Calculated conversion across the mold thickness at the end of filling for several axial positions. Experimental conditions from Figure 5.10.

b)   Temperature across the mold at the same locations (adapted from Castro and Macosko, 1982).

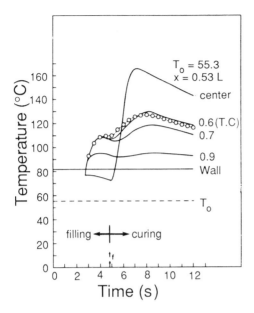

Figure 5.12 Comparison of calculated and measured temperature vs time for the experiment of Figure 5.9 (adapted from Castro and Macosko, 1982).

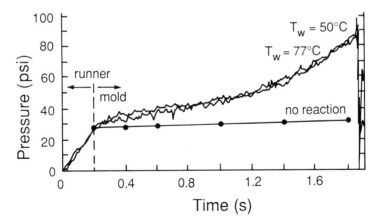

Figure 5.13 Pressure vs filling time at the entrance to a long runner (9.5mm dia) which feeds the end of a rectangular cavity 457 x 254 x 3.2mm with an ethylene glycol formulation similar to F-4, Table 2.4 at Q=220g/s, T_o=49°C. Two different mold wall temperatures: 50 and 77°C. At T_w=154° the slope was slightly higher. Straight line calculated assuming no reaction during filling (adapted from Vespoli and Alberino, 1985).

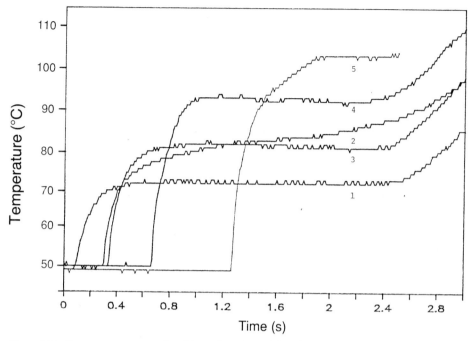

Figure 5.14   Temperature at several positions along the mold cavity of Figure 5.13 during filling with a polyurea formulation, F-7 Table 2.4 (adapted from Vespoli et al., 1985).

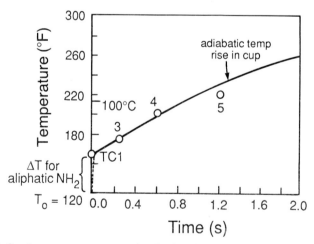

Figure 5.15   Steady state temperatures vs time for flow front to reach the thermocouple from Figure 5.14 compared to the adiabatic temperature rise for the same formulation. The rapid rise at time zero is due to the almost instantaneous reaction of the aliphatic amine, see Figure 2.12.

The adiabatic gel time can be calculated by integrating the kinetic equation up to the gel conversion, α_g, under adiabatic conditions. For a second order reaction this becomes

$$t_{g,ad} = \int_0^{\alpha_g} \exp\,[E\,/\,(\Delta T_{ad}\,\alpha + T_0)]\,d\alpha/A\,(1-\alpha)^2 \qquad (5.9)$$

where $\Delta T_{ad}\alpha + T_0$ is the temperature, since under adiabatic conditions with constant heat capacity and heat of reaction $\alpha = (T-T_0)/\Delta T_{ad}$. Gel conversion can be calculated from branching theory for crosslinking formulations, e.g. Equation 2.11. Adiabatic viscometry, as described in Chapter 2.7, can be used to test these predictions and to determine values for phase separating formulations. It is important to note that α_g may not be constant with temperature for phase separating systems. It is likely to increase especially near the phase transition temperature.

To test whether filling is adiabatic we can compare $t_{g,ad}$ to a characteristic time for conduction

$$t_{cond} = H^2\,/\,4a \qquad (5.10)$$

where a is the thermal diffusivity, typically $10^{-7}m^2/s$ for many polymers. Thus for a typical mold thickness of 3mm or greater, $t_{cond} > 20$ s. For the cases of slow reaction or very thin molds gelation will be slower than adiabatic (unless the mold is very hot) and Equation 5.8 will serve as a conservative criterion.

The simple mold filling criterion of Equation 5.8 can be tested with the data of Figure 5.7. Castro et al. (1980) report $t_{g,ad} = 2s$ for their RIM 2200 formulation at $T_0 = 50°C$. With the reaction kinetics, Equation 5.9, integrated numerically, gives the solid line in Figure 5.16. This nicely divides the completely filled plaques from the short shot.

The same approach was used to construct Figure 5.17. For this system Castro and Macosko (1982) give $t_{g,ad} \approx 5s$ at $54°C$. Again the adiabatic gel time separates the complete plaques (e.g. Figure 5.6) from the short shots (e.g. Figures 5.9 and 10). Note that both the thin plaques, 1.6mm, and the thicker ones obey the same criterion, again evidence for the assumption of adiabatic filling. If heat transfer to a hot wall (80°C) is included during filling there is some influence particularly at low T_0 as shown by the solid circles in Figure 5.17.

Figure 5.18 shows mold filling results for several initial temperatures, T_0, on a crosslinking urethane, F-1, Table 2.4 (Manas-Zloczower and Macosko, 1988). For this

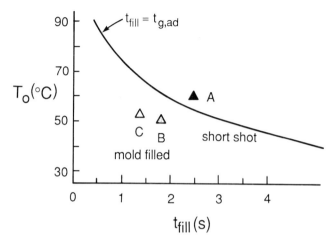

Figure 5.16 Initial temperature vs mold filling time for the experiments of Castro et al. (1980), Figure 5.7. Δ are completely filled plaques and ▲ are short shots, incompletely filled. T_0 is based on the mix average temperature of the two reactants. Line drawn from Equation 5.9 using kinetic constants from adiabatic measurements and $\alpha_g = 0.65$ from the adiabatic gel time of 2s at T = 50°.

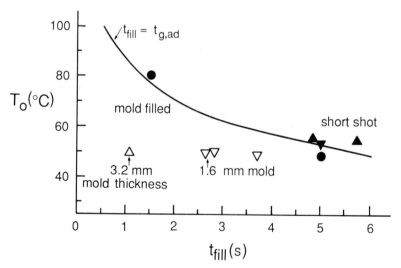

Figure 5.17 Initial temperature vs mold filling time replotted from Castro and Macosko (1982). Same notation as Figure 5.16. $\alpha_g=0.85$ from $t_{g,ad}\approx5s$ at $T_0= 54$°C. ● onset of rapid pressure rise calculated with the complete filling model including heat transfer to walls, $T_w = 80$°C.

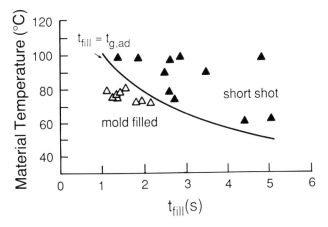

Figure 5.18   Initial temperature vs mold filling time for a crosslinking polyurethane, α_g=0.71 (replotted from Manas-Zloczower and Macosko, 1988).

formulation the theoretical gel conversion is 0.71. When combined with the reaction kinetics through Equation 5.9 it predicts well the onset of incomplete mold filling.

5.5 Moldability Diagrams for Filling

From Figures 5.16 to 18 we can see that gelation sets an upper bound to the mold filling time. In order to fill a given mold the engineer has only two variables which he can manipulate: t_g and t_f. Gel time can be increased by decreasing reactivity, i.e. by lowering the catalyst level or the average reactant temperature. However, decreasing catalyst will increase the demold time and can cause reduction in properties in phase separating polymerizations (Chapters 2.7 and 6).

Decreasing reactant temperature can lead to problems with mixing. The results in Chapter 4 indicate a critical Reynolds number for good impingement mixing

$$Re_A = \frac{4\rho_A Q_A}{\pi \eta(T_A) d_A} > 300 \tag{4.1}$$

where viscosity typically has an Arrhenius dependence on temperature as indicated in Equation 2.43. Thus if temperature is decreased viscosity increases and Re is reduced eventually leading to poor mixing.

Since the reaction is governed by the *average* temperature and mixing depends on the viscosity of each reactant, it is possible to slow down the reaction somewhat without loss of mixing quality by decreasing the temperature of the less viscous reactant.

$$T_o = \frac{Q_A T_A + Q_B T_B}{Q_A + Q_B} \tag{5.11}$$

However, this strategy is obviously limited by the critical value for Re_B; both reactants must exceed the critical Re. Poor mixing serves as an effective lower bound for reactant temperature as illustrated in Figure 5.19. Note that the flow rate in Equation 4.1 can be replaced by mold volume over filling time, $(Q_A + Q_B) = V/t_f$.

Clearly the simplest way to avoid a short shot is to decrease t_f, but as we saw in section 5.2 fast filling leads to air entrainment. The causes of air entrainment are not so well studied, but again Reynolds number, now based on mold thickness and $\eta\,(T_o)$, seems to be an important criteria.

$$Re_f = \frac{\rho v H}{\eta(T_o)} < 30 \tag{5.12}$$

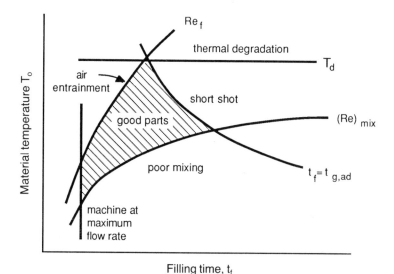

Figure 5.19 Moldability diagram to achieve a well mixed, completely filled mold without air entrainment or degradation.

The particular value of Re_f may depend on mold design (gates, inserts). Here Becker's (1979) results for flow from film gates are used. For simple molds the velocity, v, can be replaced by mold length over fill time, L/t_f. Equation 5.12 sets an effective minimum on filling time as shown in Figure 5.19. Another obvious limit on t_f, indicated in Figure 5.19, is the maximum flow rate of the machine.

For some reactants thermal stability can be important. This sets an upper limit on their storage temperatures, $T_{d,A}$ and $T_{d,B}$. The polymer produced may also be subject to thermal degradation at a temperature, T_d. The lower of these two restrictions thus sets a limit on the average reactant temperature

$$T_o < \frac{Q_A T_{d,A} + Q_B T_{d,B}}{Q_A + Q_B} \qquad (5.13)$$

or

$$T_o < T_d - \Delta T_{ad} \qquad (5.14)$$

All these criteria can be summarized in a moldability diagram for the mixing and filling steps of RIM similar to those diagrams used in TIM (Rubin, 1972). Equation 5.8 for gelation, 4.1 for mixing, 5.12 for air entrainment and 5.13 or 5.14 for degradation define the window of moldability for a particular mold, RIM machine and chemical system (Manas-Zloczower and Macosko, 1986 and 1987). This is shown schematically in Figure 5.19. Such a diagram can be readily constructed for a new formulation or mold from the mold dimensions, initial reactant viscosities as a function of temperature, the reaction kinetics and the adiabatic gel time or conversion.

This has been done for the crosslinking polyurethane formulation, F-1, in Figure 5.20. We see that the line $Re_m = 300$ correctly divides the experimentally observed good and poor mixing results. The plaques were not tested for air entrainment. Using Equation 5.14 degradation is not expected below $90\,^{\circ}C$.

Moldability diagrams do not give the optimum operating point. This is determined by additional variables like mold temperature and detailed shape as well as the curing step discussed in the next chapter. Ultimately the best operating conditions must be verified with the actual equipment and reactants. However, moldability diagrams are extremely valuable for finding the proper operating window of a new mold or a new formulation. They identify critical variables and help to sharpen thinking. They can answer such questions as: What initial temperature should be used? Is the machine flow rate adequate to fill the mold? Can the proposed formulation be mixed and flow into the mold?

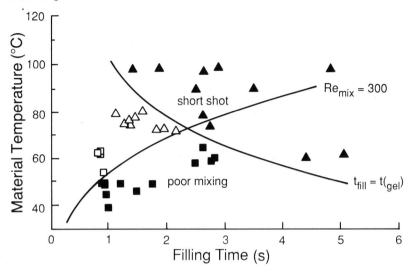

Figure 5.20   Moldability diagram for a crosslinking polyurethane (F-1, Table 2.4) compared to mold observations of complete filling △ vs short shot ▲ and good □ vs poor ■ mixing (replotted from
Manas-Zloczower and Macosko, 1988).

5.6 Filling Complex Molds

The studies described above are for filling of simple rectangular molds.   There are
some additional considerations when dealing with flow through gates into complex molds
and especially around inserts. The major problem is again the low viscosity of RIM reactants and air entrainment.

Frequently, reactants enter the mold through an aftermixer such as the one shown in
Figure 4.4. Müller et al. (1984) have shown that the "harp" or reimpingement type, Figure
4.4a, is particularly prone to air entrainment. The main value of these aftermixers is to provide back pressure to insure a full mixing chamber. Current practice accomplishes the
same function with simpler flow diverters like Figure 4.4b. No runner restriction at all
may be needed if a throttling system like Figure 4.5 is built into the mixhead.

Typical RIM parts are relatively thin and flat, thus flow from the mixhead must be
spread out. This is accomplished by one of three basic gate designs: direct, fan or dam
type. A typical direct or radial gate is shown in Figure 5.21. In visualization studies Begemann et al. (1986) found that if the mold thickness, H, near the entrance is $\leq d/8$ there were
no large bubbles entrained. This criterion is, however, difficult to satisfy for typical mold
thicknesses, $H \geq 3mm$, since it requires a very fat runner, $d \geq 24mm$. Typical mixheads
have an inside diameter of 10 -16 mm. If runner diameters smaller than 8H are used, it is

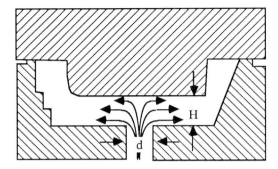

Figure 5.21 Side view of a direct fill or radial gate (adapted from Becker, 1979).

necessary to operate at Re < 100 where Re is based on H and the radial velocity at the mold entrance (Mobay, 1978; Becker, 1979). For molds in which the gate area will be removed later, local restrictions can be built in to satisfy the d/8 requirement (Mobay, 1977, 1982 and 1978; Begemann et al., 1986).

A direct gate requires mounting the mixhead near the center of the mold. This is difficult to do in most presses. Mixheads are much more frequently fitted to the mold parting line and thus fan or dam gates are the rule for RIM. Figures 4.4a and 5.22 show typical fan gate designs. Becker (1979) recommends a fan angle of < 90° to prevent jetting into the mold cavity. Müller et al. (1985) and Begemann et al. (1986) report 20° is best to avoid detachment and air entrainment. Cross sectional area should be constant or slightly decreasing through the runner and fan. Mobay (1978; Becker, 1979) recommend a land length $l_e \geq 4s$ with s=1-2mm. As pointed out in Section 5.2 they also recommend $v \leq 1.5$ m/s in this region to avoid air entrainment.

Because fan gates require a small divergence angle they take a lot of platen space to spread out the flow to the mold width. Dam gates are more difficult to machine but are much more compact as indicated in Figures 4.4b, 5.23 and 5.24. Begemann et al. (1986; also Müller et al., 1985; Maier and Menges, 1986) found that air entrainment occurs at the junction between the runner and the flow spreader in designs like Figure 4.4b. They recommend the converging design shown in 5.23. The rod type design in Figure 5.24 allows the mixhead to be mounted on the side of the mold.

All mold design studies recommend that the flow from the gate into the mold cavity be directed along the mold wall. Because of the low viscosity of RIM liquids the orientation of the mold with respect to gravity is important. This is illustrated for a lower density foaming system in Figure 5.25. To get the proper orientation tiltable mold clamps are

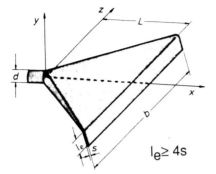

Figure 5.22   Fan gate design (from Müller et al, 1985).

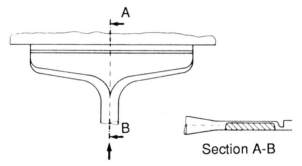

Section A-B

Figure 5.23   Dam gate with converging transition from runner to distribution channels to reduce air entrainment (from Begemann et al., 1986).

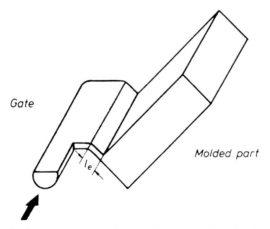

Figure 5.24   Rod or sprue gate permits mounting of the mixhead on the side of the mold (from Müller et al., 1985).

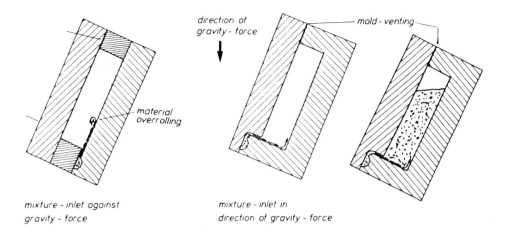

Figure 5.25 Proper orientation of mold cavity and gate with respect to gravity can reduce air entrainment problems during filling and during the mold packing step (from Müller et al., 1985).

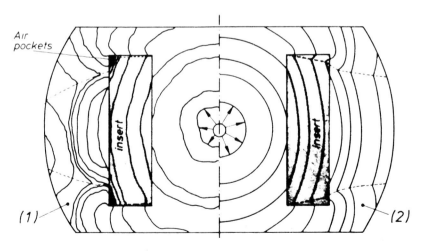

Figure 5.26 Experimental (1) and calculated (2) flow front positions for flow into a mold with a thin insert section. Air pockets and knit lines form near the rear corners of the inserts (from Müller et al., 1985).

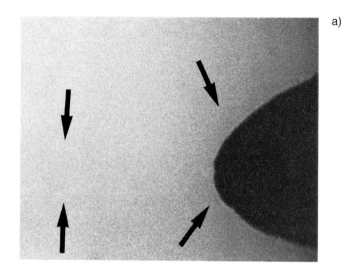

a)

b)

Figure 5.27 Photographs of two flow fronts coming together, Bayflex 110 with 50 vol % dissolved air, $T_o = 50°C$, $T_w = 60°C$, $Q = 330$ cm^3/s a) 2 mm thick mold, 0.45s after filling started, knit line is not visible, b) 4 mm thick, 0.8s after filling. In b) the knit line is clearly visible. The region shown in each photo is about 3 x 4 mm. Air pockets form at the rear corners of the inserts. These can be eliminated if the filling pressure or subsequent foaming pressure is strong enough. If not, additional vents must be placed in these locations (from Maier, 1987).

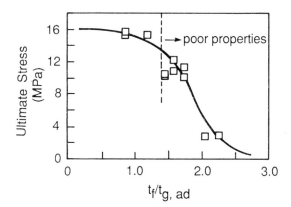

Figure 5.28 Tensile strength across a knit line as a function of filling time over adiabatic gel time. Tensile bars cut 10 mm behind a cylindrical insert. Mold: 6.4 mm thick, T_w = 48°C, T_o = 53-61°C, RIM 2200 replotted from (replotted from Castro et al., 1984).

often used in RIM (see Chapter 6). However, for high density, thin parts simple horizontal orientation can be used.

Because pressure is low during filling and during the foaming step which is used for packing, it is difficult to remove air from pockets and corners of molds. Vent location becomes extremely important. Vents should be at the highest points in the mold and in the regions where flow fronts come together. Figure 5.26 shows experimental and calculated filling profiles for flow around a thin insert section. Air pockets can form behind inserts. These can be eliminated if the filling pressure or subsequent foaming pressure is strong enough. If not, additional vents must be placed in these locations.

Figure 5.27 shows high speed photographs during knit line formation. Both are from the same material, but Figure 5.27a is 0.45s after filling starts and 5.27b is 0.8s. In the older material it appears that air bubbles are bursting at the flow front and being stretched out across the mold surface. In the older sample bubbles have had more time to grow and the viscosity is significantly higher. Since the pressure is low especially near the front at the end of the shot, these broken bubble tracks will not heal over. Gross and Angel (1975) have discussed this problem with respect to molding of thermoplastic structural foam.

Castro et al. (1984) showed that the knit line region behind an insert can be mechanically weaker especially when the insert is near the end of a long flow path and fill

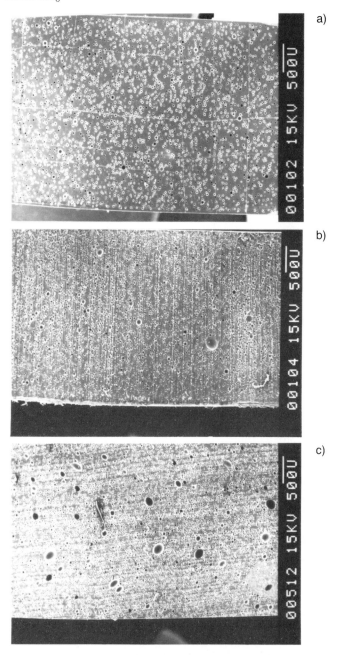

Figure 5.29 Scanning electron micrograph from a polyurethane plaque 3.2 mm thick, 63.5 cm wide, Q = 1.72 kg/s. Top of photos are at top surface of part. a) 4 cm b) 52 cm c) 123 cm from gate, end of mold cavity (from Blake and Macosko, 1988).

time excedes $t_{g,ad}$. Their tensile strength data are shown in Figure 5.28. The data correlate well with $t_f / t_{g,ad}$, additional support for the assumption of adiabatic filling. Mechanical weakness and the appearance of blemishes near knit lines are primarily due to low density rather than to poor healing of the two flow fronts, as is the case in injection molding of viscous thermoplastic melts.

Even if all the trapped air is vent out properly as discussed in 5.2, large air bubbles can become entrained at unstable flow fronts. The small air bubbles used for packing the mold may also coalesce during filling. Evidence for this is shown in Figure 5.29. A polyurethane plaque was sectioned near the gate, near the middle of the plaque and at the end farthest from the gate. The scanning electron micrographs show a pronounced increase in large bubbles with distance away from the gate. $Re \approx 10$ during filling of this part.

Other aspects of mold design will be discussed in Chapter 6. Further details on gate design and vent type and location are available in the references cited in this section.

REFERENCES

Becker, W.E. "Reaction injection molding," van Nostrand Rheinhold: New York 1979.

Begemann, M.; Maier, U.; Müller, H.; L. Pierkes, "RIM - improved technology of the machines and selective mould design to enhance moulding quality," translated from the Conference Handbook of the 13th Inst. für Kunststoffverarbeitung Kolloquium, Aachen, Block 13, p. 465 (1986).

Blake, J.W.; Macosko, C.W. "Foaming in microcellular RIM," *J. Cellular Plast.*, submitted **1988**.

Broyer, E.; Macosko, C.W.; Critchfield, F.E.; Lawler, L.F.; "Curing and heat transfer in polyurethane reaction molding," *Poly. Eng. Sci.* **1978**, *18*, 382-387.

Castro, J.M.; Macosko, C.W.; Tackett, L.P.; Steinle, E.C.; Critchfield, F.E.; "Premature gelling in RIM," *Soc. Plast. Eng. Tech. Papers* **1980**, *26*, 423-427.

Castro, J.M.; Macosko, C.W.; "Studies of mold filling and curing in the reaction injection molding process," *AIChE Journ.* **1982**, *28*, 250-260.

Castro, J.M.; Macosko, C.W.; Perry, S.J."Viscosity changes during polymerization with phase separation," *Polym.Comm.* **1984a**, *25*, 82-87; Ibid. **1985**, *26*, 138.

Castro, J.M.; Macosko, C.W.; Critchfield, F.E. "Effect of processing conditions on premature gelling, knit line strength and physical properties for the RIM process," *J. Appl. Polym. Sci.* **1984**, *29*, 1959-1969.

Coyle, D.J.; Blake, J.W.; Macosko, C.W. "The kinematics of fountain flow in mold filling," *AIChE J.* **1987**, *33 no 7*, 1168-1177.

Domine, J.D.; Gogos, C.G. "Simulation of reactive injection molding," *Polym. Eng. Sci.* **1980**, *20*, 847.

Gross, L.H.; Angell, R.G. "Smooth surface structural foam," *Soc. Plast. Eng. Tech. Papers*, **1976**, *22*, 162.

Knipp, U. "Plastics for automobile safety bumpers," *J. Cellular Plast.* **1973**, 76-84.

Lawler, L. F., Union Carbide Corp., unpublished data (1980).

Lekakou, C.N.; Richardson, S.M. "Simulation of reacting flow during filling in reaction injection molding (RIM)," *Polym. Eng. Sci.* **1986**, *26*, 1264-1275.

Maier, U.; Menges, G. "Design of RIM molds," Soc. Plast. Eng. Tech. Papers **1986**, *44*, 1350-1355.

Maier, U. "Mould design to manufacture PUR-mouldings using the RIM-process," doctoral thesis, Inst. für Kunststoff Verarbeitung, RWTH Aachen, 1987.

Manas-Zloczower, I.; Macosko, C.W. "Moldability diagrams for reaction injection molding," *Polym. Process Eng.* **1986** *4*, 173-184.

Manas-Zloczower, I.; Macosko, C.W. "Moldability diagrams for reaction injection molding of a polyurethane crosslinking system," *Polym. Eng. Sci.* **1988**, *28*, 000.

Manas-Zloczower, I.; Blake, J.W.; Macosko, C.W. "Space-time distribution in filling a mold," *Polym. Eng. Sci.* **1987**, *27*, 1229-1235.

Manzione, L.T. "Simulation of cavity filling and curing in reaction injection molding," *Polym. Eng. Sci.* **1981**, *21*, 1234-1243.

Mobay Chemical Corp., "Bayflex polyurethane elastomer mold design manual," (1977; 1982).

Mobay Chemical Corp., "RIM/Baydur structural foam mold design manual," (1978).

Müller, H.; Mrotzek, W.; Menges, G. "Mold filling with polyurethane," in "Reaction Injection Molding," J.E. Kresta, ed., ACS Symp. Series **1985**, *270*, 237.

Müller, H.; Küper, B.; Maier, U.; Pierkes, L. "The latest on molding of polyurethanes," *Adv. Polymer Tech.*," **1984**, *5*, 257-304.

Rubin, I.I. "Injection molding - theory and practice," Wiley-Interscience, New York (1972).

Vespoli, N.P.; Alberino, L.M. "Comparison of rheological measurements during mold filling for glycol and amine extended urethane RIM systems," *J. Elast. Plast.* **1985**, *17*, 173-182.

Vespoli, N.P.; Alberino, L.M.; Peterson, A.A.; Ewen, J.H. "Mold filling studies of polyurea RIM systems," *J. Elast. Plast.*, **1986**, *18*, 159-175.

6

CURING

Let's look again at the unit operations of reaction injection molding in Figure 1.2. Chapter 3 taught us how to deliver two reactive components at the proper ratio and pressure for impingement mixing. Chapter 4 described that mixing process and Chapter 5 discussed how to get fast reacting liquids into a mold without large bubbles at a conversion less than gelation. In this chapter we will complete our study of the RIM process with the curing, demolding and finishing steps.

As Figure 3.3 indicates the curing step constitutes at least a third of a typical RIM cycle. During curing, modulus and strength must build quickly for rapid demolding but also, perhaps aided by a postcure step, build optimum final use properties. It is this delicate balance between speed and performance which means the success or failure of a RIM product.

To understand the curing step we will first analyze how conversion proceeds in the mold after filling has ended and how this in turn determines temperature profiles throughout the part. The conversion and temperature profiles govern property build up, which in turn controls demolding. These temperature and demolding criteria can be combined to construct moldability diagrams for the curing step. Moldability doesn't guarantee optimal final properties so we will examine further how process variables influence properties. We will also look briefly at some aspects of mold design and at the various finishing steps that are used in RIM.

6.1 Analysis of Conversion and Temperature Profiles

In Chapter 5 conversion and temperature profiles were calculated during filling. The momentum balance (Equation 5.1) was solved together with the energy and mole balances (Equation 5.6 and 7). After flow stops we can drop the momentum balance but we still need the temperature and conversion profiles at the end of filling as the initial conditions to start our curing calculations. Thus whenever there is significant conversion during filling it is necessary to solve the more complex filling model in order to do the curing step.

Figure 5.11 and 5.12 show that there can be considerable conversion during filling. Figure 6.1 illustrates this for a similar polyurethane system (F-3 in Table 2.6). Conversion is plotted against mold thickness and length at various time increments from filling through curing. Figure 6.1a shows the mold half filled. Figure 6.1b is just at the end of filling.

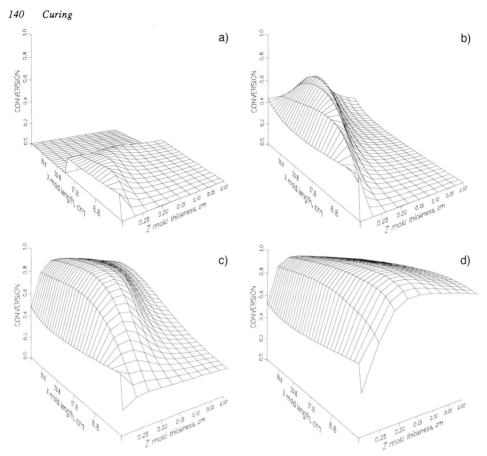

Figure 6.1   Conversion vs. mold thickness and length for polyurethane molding conditions for curve C in
Figure 5.7 (RIM 2200, F-3 in Table 2.6) at various time intervals from the filling step
through curing. a) 1.2s; b) 2.4s, end of filling; c) 3.6s; d) 4.8s (courtesy, Manuel Garcia,
1988).

Along the line of maximum residence time (see Figure 5.4) conversion is over 60%. The
gel conversion for this system is 65% and takes only 2s under adiabatic conditions. In
6.1c the nearly adiabatic reaction in the center of the mold has caught up with the maximum
residence time line wiping out gradients across the mold thickness, except near the wall.
However, although filling ended 1.2s before, there is still a strong gradient from front to
back of the mold. In the last sequence these gradients along the length of the mold have essentially disappeared and final curing is controlled by heat transfer to the wall.

However, in some cases we can neglect conversion during the filling step. The assumption holds when fill time is significantly less than the *adiabatic* gel time for mixing activated reactions:

$$t_f \leq 0.2\, t_{g,ad}\, (T_o) \qquad\qquad (6.1)$$

One example is obviously a small part where fill time is short. Another example is a slower reacting system such as nylon, inhibited dicyclopentadiene and lightly catalyzed polyure-thanes. However, very fast polyurethane and polyurea formulations will require an integrated filling and curing analysis. In Figure 6.1b $t_f = 1.2\, t_{g,ad}$.

Figure 6.2 illustrates the case of little curing during filling. It is the same polyure-thane system as in Figure 6.1 but the mold is filled in only one sixth the time, $t_f = 0.2 t_{g,ad}$. We see that conversion everywhere is less than 6% at the end of filling. Figure 6.2b shows that curing proceeds nearly independent of distance down the mold. Conversion during filling can probably be ignored and a simple one dimensional heat transfer model with reaction can be used to model curing. Figure 6.3 illustrates experimental results on a similar polyurethane for a case of fast filling. Curing begins everywhere (except close to the wall) at T_o.

For heat activated systems such as epoxy and vinyl esters fill time should be less than the *isothermal* gel time based on the mold temperature; $t_{g,iso}\,(T_w)$ should be substituted in Equation 6.1. Note that for the case of RIM 2200 $t_{g,iso}\,(65°C) = 10s$. Using this in Equation 6.1 gives t_f too long, since RIM 2200 is not a heat activated system.

Wiley and Nied (1983) have used a slightly different criteria for heat activated polymerization. They show that reaction during filling of a thermoset injection mold can be neglected if fill time is less than the characteristic reaction time

$$t_f < t_{rxn} = \frac{1}{10k\,(T_w)} \qquad\qquad (6.2)$$

where $k(T_w)$ is a first order rate constant evaluated at the mold wall temperature.

Another approach is to check for significant heat transfer during filling by comparing t_f to the Fourier conduction time

$$t_f < t_{cond} = \frac{H^2}{4a} \qquad\qquad (6.3)$$

where $a = k_{th}/\rho\, C_p$ the thermal diffusivity and H is the mold thickness. Figure 6.4 illustrates the good agreement that can be achieved between the simple curing model and experimental temperature rise on a heat activated epoxy system when filling is fast compared to reaction.

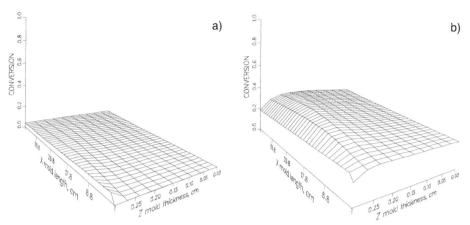

Figure 6.2   Conversion vs. mold thickness and length for the same conditions and material as Figure 6.1 except the fill time is much shorter. a) Conversion of the end of filling, 0.4s, is much less than for the slower filling case, Figure 6.1b. b) Conversion 1.2s after filling is more uniform in the x (length) direction than in Figure 6.1c (courtesy, Manuel Garcia, 1988).

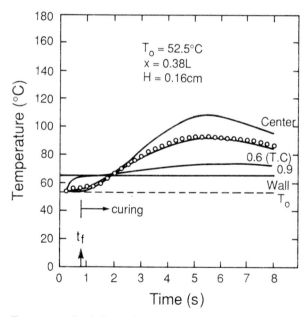

Figure 6.3   Temperature rise during curing of polyurethane (prototype BDO in Table 2.6) in a thin, 1.6 mm mold: —— calculated, ○ measured by a thermocouple in the mold cavity (adapted from Castro and Macosko, 1982).

For cases where we can ignore conversion during filling we only need to solve the unsteady energy and mole balances across the part thickness, z. Dropping the convective terms in Equations 5.6 and 5.7 gives (Estevez and Castro, 1984)

$$\frac{\partial T^*}{\partial t^*} = \frac{1}{Da} \frac{\partial^2 T^*}{\partial z^{*2}} + k^* (1 - \alpha)^2 \tag{6.4}$$

$$\frac{\partial \alpha}{\partial t^*} = k^* (1 - \alpha)^2 \tag{6.5}$$

where Da is the Damköhler number defined below. Temperature is made dimensionless by the initial temperature and the adiabatic rise, $T^* = (T - T_0)/\Delta T_{ad}$; time is made dimensionless by the reaction rate, $t^* = k (T_0)C_0 t$; position across the mold by half the thickness, $z^* = 2z/H$ and the rate by the initial rate $k^* = k(T)/k(T_0) = \exp[-E_a(T^{-1} - T_0^{-1})/R]$. Here as in Chapter 5, we have assumed second order kinetics but any other simple kinetic expression which depends on the overall conversion of the functional groups, α, can be introduced into Equation 6.4 and 5.

As discussed, we ignore the filling step and assume that initially the mold is filled everywhere with liquid at the average tank temperature and no conversion

$$T (0, z) = T_o \quad \text{and} \quad \alpha (0, z) = 0 \tag{6.6}$$

The boundary conditions are symmetry along the center of the cavity

$$\frac{\partial T}{\partial z} = 0 \tag{6.7}$$

and isothermal walls

$$T (t, z^* = 1) = T_w \tag{6.8}$$

Later we will consider non-isothermal molds.

In solving Equations 6.4 and 5 the Damköhler number is the key parameter.

$$Da = \frac{\text{heat produced by reaction}}{\text{heat transfer by conduction}} = \frac{C_o k (T_o)}{4a/H^2} \tag{6.9}$$

When Da is large reaction heat dominates and the mold becomes adiabatic. When Da is small heat transfer dominates and Equation 6.4 is uncoupled from 6.5. Mold temperatures can be found from the usual transient heat transfer in a slab.

The role of the Damköhler number is illustrated in Figure 6.5 where the maximum temperature in the center of a part is plotted against Da. At low Da, T_{max} is just T_w. As Da increases above unity T_{max} goes to $T_0 + \Delta T_{ad}$.

It is helpful to illustrate this plot with an example. Let's use it to predict the maximum temperature in Figure 6.3. From that figure $T_0 = 53°C$, $T_w = 65°C$ and H = 0.16 cm. From Table 2.6 for prototype BDO formulation $C_0 = 2.6$ mol/L, catalyst concentration is 0.075 wt%, $k(53°) = 0.02$ L/mol·s and $\Delta T_{ad} = 115°$. Typically a $\approx 10^{-3}$ cm^2/s. Using these parameters gives $T_w^* = T_w/\Delta T_{ad} = 2.93$ and Da = 0.33. Reading Figure 6.5 gives $T_{max}^* = 0.38$ or $T_{max} = 100°C$. This is close to the maximum temperature (~107°C) at the center position given in Figure 6.3.

There are a number of other references with Damköhler plots for other kinetics, activation energies and wall temperatures (Perry et al., 1985; Williams et al., 1985). For studies with a particular resin (E, n and A* fixed) and a given initial or wall temperature such plots can be helpful. However, in general there are too many parameters to make them very useful.

It is usually more helpful to do a numerical solution. It is quite straight forward to solve Equations 6.4 and 5 via finite differences. An implicit difference equation works well with 10 to 20 increments in z*. Time steps are adjusted to keep changes in conversion small $0.01 < \Delta\alpha < 0.03$ (Broyer and Macosko, 1976). The difference equations can be solved with a microcomputer. With such a program temperature and conversion vs. time profiles during simple curing can readily be generated for a wide range of materials and boundary conditions. Results compare well with experiments (e.g. Figures 6.3, 6.4; also Broyer et al., 1978; Vespoli and Alberino, 1985).

It is not always correct to assume an isothermal mold wall. Heat is often generated so rapidly by the reaction that it cannot be carried away from the mold walls. Lee and Macosko (1980) included heat transfer through the walls and to cooling water in their analysis. Figure 6.6 shows that wall temperature can increase nearly 20°C during a typical polyurethane cure. For a 25 mm thick steel wall the rise was about 7°. Lee and Macosko give experimental confirmation of these calculations. Barone and Caulk (1985) also show that there can be gradients *across* the mold surface. In production it is typically not possible to conduct away all the heat from the first cycle before the second begins. Thus the

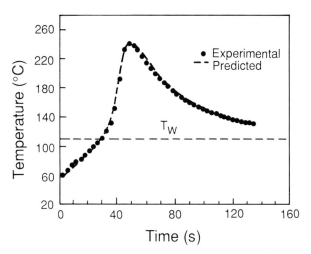

Figure 6.4   Temperature rise at the center of a thick, 4.8 mm mold during curing of an epoxy RIM formulation (from Manzione and Osinksi, 1983).

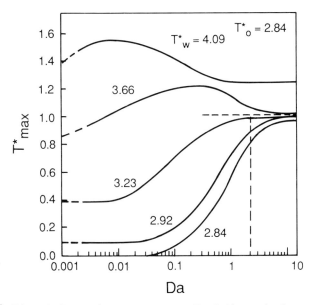

Figure 6.5   Dimensionless maximum temperature vs. Damköhler number for several values of dimensionless wall temperature. Temperatures are made dimensionless by the adiabatic temperature rise $\Delta T_{ad} = 115$ K. Thus $T_0 = 55°C$ and T_w ranges from $55°$ to $200°C$. $T_{max}^* = T_{max} - T_0) / \Delta T_{ad}$. Kinetics are second order with parameters similar to prototype BDO in Table 2.6. $E^* = E_a/R\Delta T_{ad} = 60$ (adapted from Estevez and Castro, 1984).

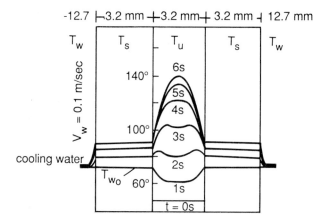

Figure 6.6   Temperature development during curing in a thin walled steel mold for a fast reacting polyurethane; v_w is the cooling water velocity. Water cooling is not sufficient to prevent temperature rise at the wall (adapted from Lee and Macosko, 1980).

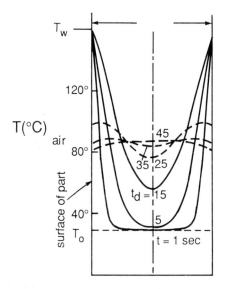

Figure 6.7   Temperature across part during a heat activated silicone rubber cure. Note that the temperatures continue to rise (dashed lines) in the center after ejection which occurs after 15s (adapted from Macosko and Lee, 1984).

mold temperature typically rises to a higher value after several cycles. In production it is only necessary to heat molds for the first few cycles. In steady operation the role of the heat transfer fluid is to remove reaction heat.

Curing can continue after the part is demolded. Figure 6.7 illustrates this for the heat activated curing of a silicone rubber with low exotherm. After demolding heat transfer at the part surface is controlled by natural convection of air. Heat inside the part near the wall is also transferred toward the cooler center. Heat transfer during curing can also decrease if the part shrinks away from the mold wall.

6.2 Mold Packing

During the curing step the part shrinks due to polymerization and to cooling. To compensate for this shrinkage foaming is used. This is a major difference between RIM and thermoplastic injection molding (TIM) where high melt pressure is used to pack the mold with extra melt to compensate for shrinkage. As discussed in Chapter 1, for large parts this high melt pressure means much larger clamp forces.

In RIM dry air or nitrogen is loaded into one or both of the reactants (see Chapter 3). In the metering cylinders the gas is compressed, bubbles collapse and dissolve. Bubbles nucleate again as pressure drops suddenly in the turbulent flow in the mixhead. They grow as they flow into the mold. Figure 6.8 summarizes the steps that a bubble goes through during the RIM process. In Chapter 5 some of the problems of coalescence and air entrainment along the moving front were described.

At the end of the filling step the bubbles continue to grow due to diffusion of dissolved gas and to expansion of the gas from the rising temperature. The froth expands to fill out the mold and then starts flowing out the vents. When material gels in the vents flow stops, the bubbles can no longer expand and mold pressure rises. This fixes bubble size, typically 30 μm diameter (Kamal et al., 1986; Blake and Macosko, 1988). Soon gelation occurs in the mold cavity. The polymer solidifies and can no longer transmit pressure to the transducer. Measured pressure reaches a maximum and then falls as the polymer cools. It can even become "negative" as polymer shrinking away from the wall pulls a vacuum.

These stages in mold pressure are shown in Figure 6.9. Similar curves have been reported by Begemann et al. (1986) on polyurea-urethanes. We see in Figure 6.9 that reducing the vent cross sectional area greatly increases the maximum pressure rise. Note that the time to the maximum is the same in each case, 8-9s after end of injection. This time corresponds to about 98% conversion in the center and 65% at $z^* = 0.9$, based on calcula-

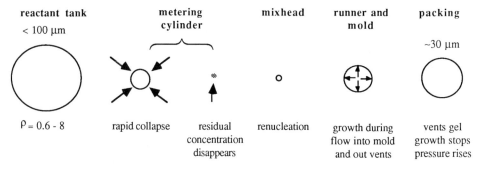

reactant tank	metering cylinder	mixhead	runner and mold	packing
< 100 μm				~30 μm
ρ = 0.6 - 8	rapid collapse	residual concentration disappears	renucleation	growth during flow into mold and out vents	vents gel growth stops pressure rises

Figure 6.8   The fate of a bubble as it goes through the RIM process (from Blake and Macosko, 1988).

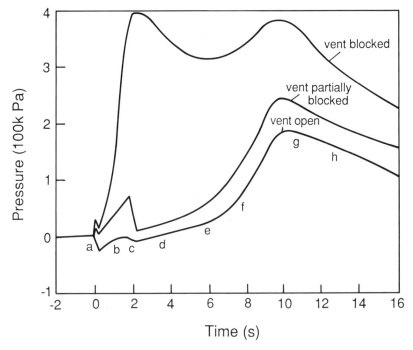

Figure 6.9   Mold pressure vs. time during filling and curing of a thick plaque (305 x 457 x 6.35 mm) for a high modulus, ethylene glycol-based polyurethane formulation. a) Transient due to temperature change as fluid crosses transducer. b) Pressure rise during mold filling. c) Drop at end of filling. Sometimes there is an extra spike here has the mixhead ram moves forward. d) Bubble growth to fill out cavity. e) Pressure rise during flow through vents. f) Vent gelation. Pressure rises in bubbles due to exotherm and diffusion. g) Gelation near mold wall. h) Solidification and cooling. The three curves are for open, partially blocked and totally blocked air vent (adapted from Blake, 1987).

tions from kinetic data and Equations 6.4 and 5. Thus the time to the pressure maximum corresponds to gelation of polymer *near the wall* (Blake and Macosko, 1988).

Some time after the pressure maximum the part can be demolded. If demolding is done too soon, especially on thick parts, "postblowing" can occur. If the modulus of the polymer is not high enough the pressure inside the gas cells can cause thick sections to expand after the constraining walls of the mold are removed. This problem is of greatest concern in sections over 6 mm thickness. Another cause of post blowing in polyurethane and urea systems is delayed reaction of excess isocyanate with water. In thick sections the by-product CO_2 cannot escape rapidly enough. A more rapid buildup of modulus reduces this problem. Demolding problems are discussed further in the next section.

Clearly processing conditions affect final part density. The biggest influence is the amount of gas loading in the reactants. Figure 6.10 shows the effect of tank specific gravity on average part density. More gas loading (lower resin tank specific gravity) gives lower part density. A similar trend is observed with level of mold filling. An underfilled mold leads to lower density since the gas bubbles must expand further to fill the mold before gelation (Okuda, 1981; Blake, 1987).

Density is not constant but decreases from the gate to the vent, typically by 10%. Bubbles near the front have more time to grow and are subject to lower pressure than near the gate. Density gradients in the flow direction may also result from bubble coalescence near the front (see Figure 5.29). Wall temperature seems to have the greatest influence on these gradients. As Figure 6.11 shows a hotter mold leads to steeper gradients. The hotter mold causes more reaction near the front at the end of filling. This results in lower density at the end of the mold but also leads to earlier gelation which can cause higher pressure and thus higher density near the gate.

There is also some gradient across the thickness of the part. A dense surface is well known and desired in low density RIM and integral skin foam (Oertel, 1985). For high density RIM Blake and Macosko (1988) report that bubbles are also slightly smaller and perhaps less numerous near the mold wall (see Figure 5.29a). However, we should note that in one study Kamal et al. (1986) report more bubbles near the wall. Again temperature gradients, i.e. hotter material in the center, cooler at the wall, seem to be the main source of the bubble size gradient.

6.3 Demolding

The minimum that the curing step must accomplish is to develop sufficient properties so that the part can be removed from the mold. At some point after the pressure maxi-

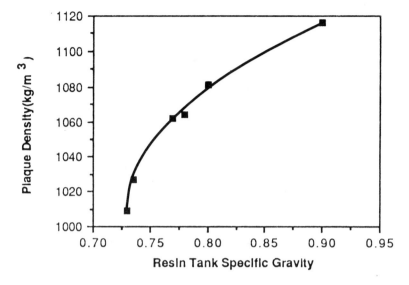

Figure 6.10   Effect of gas loading in polyol on average part density for a low modulus ethylene glycol formulation. Plaque mold, 1270 x 635 x 3.2 mm (from Blake and Macosko, 1988).

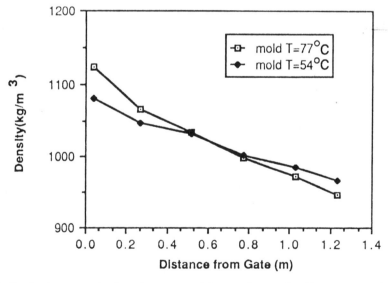

Figure 6.11   Density decreases going from the gate toward the vent. Increased mold temperature enhances
that gradient. Same mold and material as in Figure 6.10 (adapted from Blake, 1987).

mum (Figure 6.9) it should be possible to remove the part, but when? What properties must the polymer have to enable demolding? To answer these questions it is helpful to divide demolding into three steps: mold release, part removal, and warpage (Figures 6.12 - 6.14).

The first step in demolding is mold opening; one half of the mold must release from the surface of the part. This is shown schematically in Figure 6.12. Release of the part from the mold surface requires that the adhesive force between mold and polymer be less than the strength of the polymer. Since RIM parts are often large, if adhesion is too strong it may be impossible to even open the mold! Typically, however, with strong adhesion something tears, either a surface skin of polymer or the flash around the edges of the part.

Clearly adhesion can be reduced by spraying mold release agents on the mold surfaces. Stearates and other fatty ester soaps or waxes are preferred over silicones since the latter interfere with most painting systems. But external release agents are undesirable in general because they require time to apply, and they tend to build up in mold corners, necessitating time consuming cleanup (Cekoric et al., 1983; Meyer, 1985). As discussed in Chapter 2.4, today most urethane and urea formulations contain internal mold release agents. The most common is zinc stearate plus a fatty acid at 1-2% concentration in the system. Silicone compounds have also proven successful but they have problems deactivating tin catalysts (Plevyak and Sobreski, 1985). Despite the use of internal release, external agents must still be sprayed on every 10 to 30 shots, depending on part complexity. The entire mold is cleaned and recoated with external release after every eight hour shift (200-300 parts). Simple parts may require no external release agents. Polyureas are claimed to release better than polyurethanes (Dominguez et al., 1987). Dicyclopentadiene and especially nylon RIM formulations also show very good mold release characteristics.

Evaluation of internal mold release agents is difficult. Typically performance is based on how many "clean" releases of a complex production part can be achieved between external sprays. More quantitative evaluation has been tried by measuring the press opening force. Willkomm et al. (1989) have measured shear release forces on a simple concentric cylinder mold. They also showed that break away torque after curing in a parallel plate rheometer gave a good correlation with molding experience. Their results indicate that a release agent must reduce the break away shear stress at the polymer metal interface to below 50 kPa for a clean separation. This is considerably lower than the pressure needed to hold the mold closed (Figure 6.9) and thus does not present a limit to press hydraulics.

After the two mold halves separate it is still necessary to get the part out of the cavity. For higher modulus formulations knock out pins can be used but large flexible parts

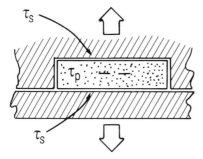

Figure 6.12 Mold release. The stress, τ_s, to release the surface of the part from the mold surface must be less than that which will tear apart the polymer, τ_p, and less than that which the press can exert, $\tau = F/A$.

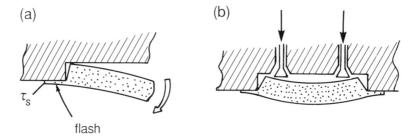

Figure 6.13   Part removal. a) large soft, elastomeric materials are usually pulled out. Failures often occur when the thin flash material around the edges is not strong enough to come off with the part but sticks to the mold and tears. b) Rigid RIM polymers are ejected with pins. The polymer must be strong enough that the pins do not go through and stiff enough that a few pins can push out the entire part.

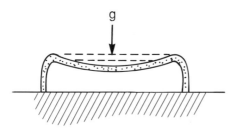

Figure 6.14 Dimensional stability. After ejection the part must hold its shape under gravity stresses as it cools down. Cooling forms are often used for large parts.

Figure 6.15 Removal of automotive fascia from mold (from Marty Cornell photos courtesy of Dow Chemical).

are often peeled from the cavity by hand or by robot. This is shown schematically in Figure 6.13 and illustrated for production of a fascia in Figure 6.15.

To come out in one piece the part must have developed enough "green" strength to survive considerable bending stresses. The most susceptible to these stresses is the flash on the edges of the part. If this flash tears off it is left on the mold land or may fall into the cavity. This can ruin the next part. Bending stresses during removal also can cause surface cracks if the part is undercured.

The most common test for sufficient bending strength is to fold over a corner of the part 180° immediately upon demolding. If the corner breaks or cracks or even if it survives the test but leaves a crease mark the part is rejected. To improve its green strength on demolding the part may be a) left in the mold longer, b) the mold heated or c) the chemistry changed, for example by increasing the catalyst level.

Figure 6.16 shows some results of the bend test for a crosslinked urethane. Immediately upon demolding the corner of a 3 mm thick plaque was bent 180° and punctured along the bend. If the puncture grew the plaque was designated torn. This was done for a series of mold temperatures and times. From Chapter 2.6 we expect crosslink density to increase directly with conversion. For elastomers, tear resistance correlates with crosslink density (Gent and Tobias, 1982). Thus a line of constant conversion should separate materials with good and poor tear resistance. Using reaction kinetics for this crosslinking ure-thane (F-1 Table 2.6) and the energy balance, Equations 6.4 and 5, we can calculate the conversion in the mold. The greatest stress during bending is at the surface so conversion there should control tear strength. In Figure 6.16 we see that the time to 90% conversion calculated at the mold surface correctly predicts when parts have built sufficient tear strength for demolding. Over the temperature range studied tear strength seems to only depend on conversion and thus crosslink density and not temperature.

The third step in demolding is holding part shape during cooling to prevent warp-age (see Figure 6.14). Gravity loads can be considerable so modulus must be high enough that the part hold its shape. If it deflects this will be frozen into the part. To pre-vent warpage large parts are often placed on support frames immediately after demolding. This is an extra step which could be avoided if the polymer builds modulus quickly enough in the mold.

The problem of warpage can be illustrated with a liquid silicone elastomer similar to the one studied by Macosko and Lee (1984). Demolding measurements are given in Figure 6.17. The silicone was premixed and injected at room temperature with a thermo-plastic injection molding machine into a hot mold. The cavity was a 15 x 15 cm plaque of varying thickness with a 130° fan gate. As indicated in Figure 6.7 the polymerization is heat activated and not very exothermic so it cures from the mold walls toward the center. The surface cures quickly so it is possible for parts to be ejected before the center is gelled. Gelation occurs during cool down and if parts are even slightly bent during cooling they will be permanently distorted. However, if they are held in the mold until the center gels there is no problem with distortion.

To determine when to demold distortion free parts Kemp (1985) pressed his thumb (well insulated) against the center of each plaque immediately upon demolding. He deter-mined the cure time required to just prevent a permanent thumbprint on the part. The gel conversion for this formulation is 0.5. Using DSC data similar to that of Macosko and Lee (1984), Batch (1989) calculated the time to 50% conversion in the center of plaques of varying thickness. Agreement with the demolding measurements, as shown in Figure 6.17, is excellent.

The above two demolding studies follow the general strategy for calculating de-mold time. This strategy is shown schematically in Figure 6.18. The energy balance and reaction kinetics are solved together to calculate conversion and temperature profiles. The important input parameters needed to solve these equations like mold temperature, thermal diffusivity, mold thickness, initial concentration and kinetic constants are indicated in Fig-

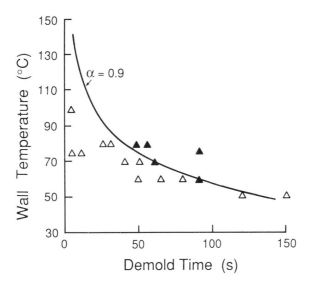

Figure 6.16 Demolding performance of a crosslinked polyurethane (Formulation F-1 Table 2.6). **Δ**- puncture on bend propagates; **▲** - puncture did not propagate; ——— $\alpha = 0.90$ at the mold surface, calculated from kinetic data (adapted from Manas-Zloczower and Macosko, 1988).

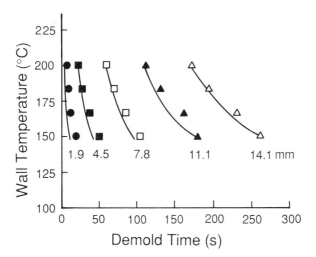

Figure 6.17 Demold times as a function of cavity thickness and mold temperature (Kemp, 1985). Solid lines are $\alpha = 0.5$, the gel conversion, at the center of the mold calculated from DSC kinetic data (Batch, 1989).

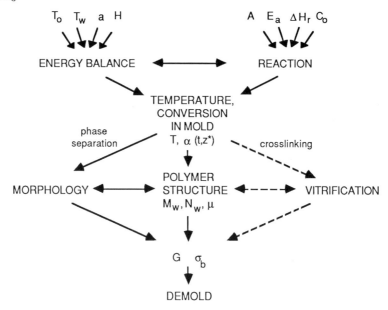

Figure 6.18 Schematic diagram for calculating stiffness (G) and strength (σ_b) which in turn control demold time. For segmented block copolymers morphology development due to phase separation plays a major role in property development. For crosslinking systems vitrification can
be important.

ure 6.18. Conversion alone generally determines polymer structure, molecular weight,
crosslink density, block copolymer sequence length. For crosslinked polymers which will
be used below their glass transition, vitrification during the curing step can be critical to
eliminate warpage (Figure 6.14). However, the primary determinant of properties is crosslink density. Thus we can usually correlate demold time simply to the time to reach a certain conversion as was done in Figures 6.16 and 17.

However, for block copolymers like polyurethanes and polyureas the development
of two phase morphology controls mechanical properties. In particular at low mold temperatures premature phase separation can cut off the polymer chain growth, giving a brittle
part. At high temperature reaction and polymer structure buildup is fast but phase separation kinetics are slow. The combined influence of phase separation and reaction on demolding is illustrated in Figures 6.19-22 for a butane diol based polyurethane.

In Figure 6.19 the lines are from infrared data (Figure 2.24b). The solid line is the
time to reach 70% conversion at different temperatures. The dashed line is the time to the
maximum in free carbonyl groups. This maximum has been identified as the beginning of
phase separation.

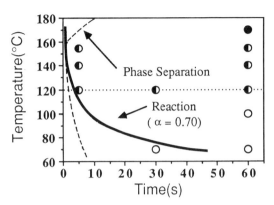

Figure 6.19   Mold surface quality after corner bend test compared with FT-IR data from Figure 2.24b,   a 48% hard segment polyurethane, 0.02% DBTL. At 120°C the onset of phase separation occurs at about 70% conversion which correlates with a transition from flaky, powdery surface appearance upon bending. ○ flaky surface; ◑ crease; ● smooth surface. ⊗ indicates sample used for GPC results in Figure 6.20 (from Yang, 1987).

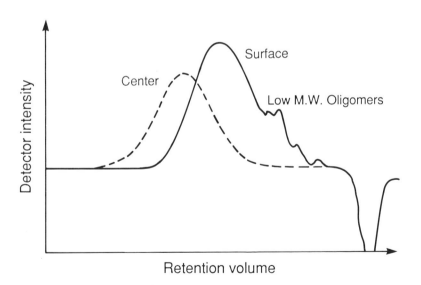

Figure 6.20   GPC chromatograms from 70°C, 60s demold sample in Figure 6.19. The solid line indicates a sample taken from the surface ($0.7 \leq z^* \leq 0.1$) and the dashed line from the center portion of the plaques (from Yang, 1987).

The points in Figure 6.19 are qualitative observations of demolding a RIM plaque 3 mm thick. Below 120°C the surface in the region of a bend was always flaky--brittle and powdery. This powdery material was found by GPC to be very low molecular weight even showing unreacted diol (see Figure 6.20). Below 160°C the surface still showed a crease. Only at 170° was the surface smooth. Tirrell et al. (1979) and Camargo and co-workers (1985) have also observed large differences between surface and center molecular weight.

Although these are only qualitative observations, they can be understood in terms of the IR data. As discussed in Chapter 2.7 low polymerization temperature and low catalyst level result in low molecular weight. Below 120°C the dotted line indicates that a second phase is forming much before the reaction can build up polymer chain length. At 120°C the maximum in free carbonyl concentration occurs at about 70% conversion. At 120° and higher temperatures the reaction is fast enough to build molecular weight before the reactants become isolated due to phase separation. Isothermal FT-IR data is valid here because, although the center of the part will be much hotter, the surface will stay close to the mold temperature.

When catalyst level was increased to 0.075 wt% the flaky surface disappeared at 100°C mold temperature. FT-IR data showed that at only 100°C this catalyst level increased reaction rate such that the free carbonyl maximum occurs at about the same time as 70% conversion. Substitution of the same equivalent weight triol for the diol completely eliminated the flaky surface. Triol will slow down phase separation and increase molecular weight buildup (Yang, 1987).

In addition to a flaky surface 70°C mold temperatures also lead to failure in the bend test, as illustrated in Figure 6.21. If the part is held in the mold longer (60s), it doesn't break but only cracks slightly on the surface. If the temperature is raised to 140°C or higher the part tears in the bend test. This illustrates a fundamental difference between crosslinking and phase separating systems. At high temperature the two phases become compatible and the polymer loses its strength and stiffness. Even when triol is substituted for the diol the results are not changed significantly, as shown in Figure 6.22. The few crosslinks in the soft segments are not sufficient to give tear strength or high modulus. The triol formulation is closer to commerical polyurethane and illustrates the dominate role of phase separation on demolding properties in RIM.

Figures 6.21 and 22 show good correlation between the bending tests and the time and temperature for modulus buildup to 1 MPa (taken from Figure 2.28b). One should not generally expect a correlation between modulus and ultimate strength. Molecular weight

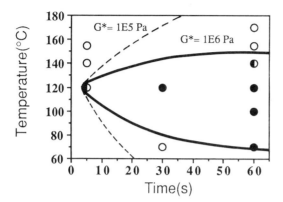

Figure 6.21 Results of bending test compared with constant modulus contours from Figure 2.28b. Complex shear modulus results: -------- time to reach 0.1MPa; —— time to 1MPa. Bending test results: ● surface crack; ◑ crack; ○ break or tear. Same formulation as Figure 6.19 (from Yang, 1987).

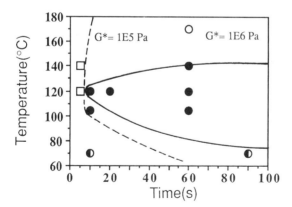

Figure 6.22 Results of bending test compared with constant modulus contours. Triol substituted for diol in formulation of Figure 6.19, 0.01 wt% DBTDL. Complex modulus: --------- time to reach 0.1 MPa; ——time to 1MPa. Bending test results: ● surface crack; ◑ crack; ○ tear; □ not gel (from Yang, 1987).

and thus extent of reaction is also critical. At low temperature modulus and apparently bending strength follow conversion; the time to $G^*=1MPa$ is about the same as the time to $\alpha=0.70$ (see Figure 6.19). However at high temperature the reaction is very fast. Modulus builds up through phase separation. In this region modulus also appears to correlate with tear strength.

On the other hand we *do* expect a direct correlation between dimensional stability or warpage and modulus growth. Yang (1987) tested the same samples described in Figures 6.21 and 22 for dimensional stability. He called a 3 mm thick plaque "warped" if immediately after demold it sagged more than 50 mm on a 200 mm overhang. The time and temperature for modulus to reach 1 MPa correlated well with the transition from warped to stiff plaques, very similar to the results in Figures 6.21 and 22.

The bend test has been applied by Willkomm et al. (1988) to polyureas, which have a serious problem with brittleness on demolding. Willkomm et al. had to raise mold temperature to 80°C to even be able to demold a plaque in one piece; 155°C was needed to get a tough part in a reasonable time, see Figure 6.23. It is important to note that after aging for a day the brittle plaques all became tough. Although they don't have data on phase separation dynamics, based on GPC results Willkomm et al. speculate that phase separation is very rapid and traps reactants. Only at higher temperatures can molecular weight build up sufficiently to obtain good "green" strength. During aging, reaction can still proceed, albeit slowly, since polyureas are totally amorphous in contrast to RIM polyurethanes. Willkomm et al. did not observe high temperature softening as with polyurethanes. Apparently the phase mixing temperature is at or above the degradation temperature in these polyureas.

Alberino and coworkers (1985) report similar brittleness problems with a trifunctional soft segment formulation similar to F-7 in Table 2.4. However, when they added about 13% high molecular weight (6000) trifunctional species to the *isocyanate* they got a significant improvement as shown by the dashed line in Figure 6.23. The reason is not fully understood. Possibly the oligomer slowed down phase separation enough to permit sufficient molecular weight growth before demolding.

6.4   Moldability

It is possible to summarize the main phenomena which must be controlled during the curing step in the form of moldability diagrams. Like the filling moldability diagrams developed in Chapter 5 these are helpful to understand the variables involved and determine if and when a part can be demolded.

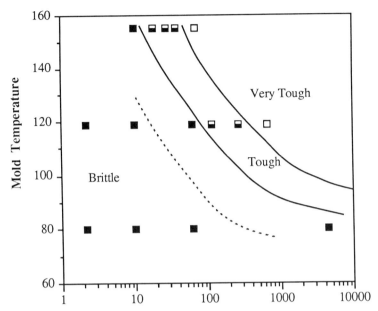

Figure 6.23   Demolding data for a 50% hard segment polyurea, 3.2 mm thick plaques. Squares are for a linear, prototype system (F-6, Table 2.4, Willkomm et al., 1988). ■ shattered in mold; ▣ cracked on bending; □ no cracking. The dashed lines shows the transition from brittle to very tough for a lightly crosslinked polyurea with 13% of a 6000 molecular weight triol blended into the isocyanate before processing (similar to F-7 Table 2.4; Alberino et al. 1985).

If we are given a part to mold by RIM what variables can we control to influence the curing stage? Mold temperature (T_w) within a limited range, is readily changeable in the process and has a large influence on curing. For some systems where there is a large difference between T_w and the initial material temperature, T_o, the latter may also play a role. But it is clearly less important than T_w. Part thickness will have a big effect on heat transfer and thus curing rate, but this has been already fixed by the part design. Formulation variables, particularly catalyst level, can have a huge effect on reaction rate, however typically these have also been specified with the part and are outside the control of the process engineer. Of course based on moldability studies recommendations are often made to alter part design or formulation chemistry. The following diagrams can greatly aid that iteration.

From the above discussion the most important variable that the process engineer can manipulate to affect curing is mold temperature. Therefore moldability diagrams can be constructed on the T_w - t_{demold} plane. We need to consider two different diagrams: one for systems which structure by crosslinking; another for structuring by phase separation.

There are two obvious limits on demolding time: a minimum time for the press to open and a maximum time. Maximum time is defined from manufacturing economics: what is the number of parts that a given mold should produce per day? This number is often set by competitive economics; for example how fast can the same part be made via TIM? As indicated in Table 1.2 and in Chapter 3, typical demold times for polyurethanes and polyureas are 30 to 60s. For lower volume production 120s or even longer may be acceptable.

Minimum and maximum demold time then define two vertical lines on the T_w - t_d plane for both crosslinking and phase separating systems:

$t_{min} \geq$ 10s (set by press opening time)

$t_{max} \leq$ 120s (set by production economics)

Crosslinking and phase separating systems also have an upper limit on mold temperature such that the total temperature throughout the part never exceeds the polymer degradation point. For example in Figure 2.2 we saw that the urethane bond begins reversing at 170°C. The depolymerization rate increases with temperature such that 200°C is probably an upper mold temperature for polyurethanes:

$T_d \leq$ 200°C (set by polymer degradation)

Another limitation on mold temperature can be a human factor. For manual part removal $T_w \leq$ 120°C.

The most important limit on demolding time is development of sufficient properties. For crosslinking elastomers Figures 6.16 and 17 indicate that demolding is controlled by either development of tear strength or complete conversion to prevent warpage. Both are governed by a critical conversion either at the part surface or the center:

$\alpha_c >$ 0.90 (surface tearing)   or

$\alpha_c >$ 0.50 (center conversion)

Once identified α_c can be readily calculated from the energy balance and reaction kinetics as outlined in Figure 6.18.

It is important to note that even if there is conversion during filling demolding is determined by the last material to enter the mold--the material near the gate . There will be essentially no conversion during filling of this region (note Figure 6.1b) and Equations 6.4

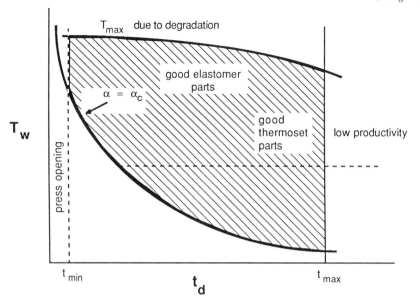

Figure 6.24   Moldability diagram for the curing step with a system which builds structure by crosslinking (adapted from Manas-Zloczower and Macosko, 1986; 1988).

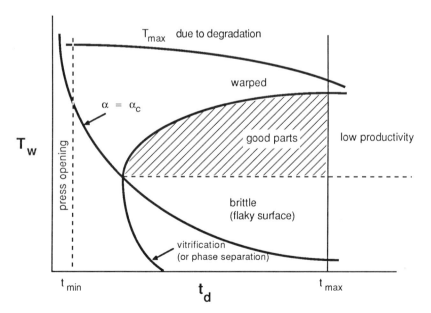

Figure 6.25   Moldability diagram for the curing step with a phase separating, linear polyurethane.

and 6.5 can be solved to determine α_c. Thus in constructing a simple moldability diagram for curing we can neglect the filling step.

These moldability criteria for elastomeric crosslinking systems are shown schematically in Figure 6.24. For thermosets an additional line is required, as indicated in Figure 6.25 the time for a surface layer of the part to reach the glass transition temperature. This depends first on conversion to high molecular weight and crosslink density and then on cooling of the part surface once maximum conversion is reached. This criteria is important for crosslinked RIM systems which will be used below their T_g like epoxies and dicyclopentadiene (see Chapter 7). If parts from these materials are demolded at $T > T_g$ they can cool and vitrify in a distorted shape. At low mold temperature vitrification can occur before sufficient reaction which leads to a brittle part. Gillham (eg Aronhime and Gillham, 1986) has reviewed extensively the role of vitrification during curing.

A similar moldability diagram can be drawn for phase separating systems. Figure 6.25 shows the same t_{min}, t_{max}, T_d and α_c lines as for crosslinking systems. Strength of the part is still controlled by buildup of polymer molecular weight which can be correlated to α_c. However, new limits are set by the phase separation kinetics. At low temperatures phase separation occurs before reaction builds high polymer, especially at the surface of the part. This leads to flaky material in the case of polyurethanes and brittleness with polyureas.

As mold temperature is increased the reaction speeds up faster than the rate of phase separation. There are two reasons for this. First, polymer molecular weight is higher at short times increasing diffusion time. Adding more catalyst has a similar effect to raising temperature (Camargo et al., 1985). Second, at high temperatures the block copolymer phases become more compatible. At a critical temperature reaction is fast enough at the mold surface to eliminate the flaky, low molecular weight material.

At higher temperatures a different phenomenon controls demolding: dimensional stability. We need to hold parts long enough in the mold until the hard phase has time to nucleate and grow to build enough stiffness to demold a rigid part. The phenomenon is similar to vitrification shown in Figure 6.24.

The demolding diagram agrees well with the results of Yang (1987) for a linear polyurethane (Figures 6.19-6.21). Light crosslinking in the soft segment eliminates the flaky surface but not the other features (Figure 6.22). Vespoli (1985) observed a transition from brittle to good to warped parts with increasing mold temperature using a lightly crosslinked MDI-EG-PTMO formulation. The polyureas never show warpage problems, presumably because the phase compatability temperature is above T_d. Nylon RIM systems

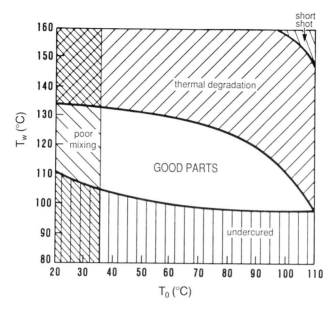

Figure 6.26 Moldability diagram combining mixing and curing step for an epoxy RIM system (4.8 mm thick, 180 μm long plaque): poor mixing, $Re \leq 250$; short shot, $\alpha \geq 0.49$ during 12s filling; degradation, $T_d \geq 265°$; undercured, $\alpha \geq 0.6$ at $t_d = 120s$ (from Manzione and Osinski, 1983).

structure by phase separation and as we will see in the next chapter nicely fall onto this demolding diagram.

If fill time and demold time are fixed at some reasonable values it is possible to combine the important limits on both filling and curing into a single diagram. This has been done by Manzione and Osinski (1983) and more recently by Kim and Kim (1987), both for epoxy systems. Figure 6.26 shows one of Manzione and Osinski's diagrams for a 4.8 mm thick, 180 mm long end-gated plaque. Fill time was fixed at 12s. Even under these slow filling conditions there is little reaction during filling except at very high initial temperature. Since filling time is fixed, flow rate through the impingement mixhead is fixed and the critical Reynolds number for good mixing is controlled by T_0 only; 35°C gives suitable mixing for the nozzles used.

Since high mold temperatures are needed for epoxies (see Chapter 7) degradation is a more important limitation than short shots. For this epoxy the adiabatic temperature rise is 233°C. Thus it is quite possible for the temperature inside the part to exceed the degradation temperature, 265°C. Demold time is fixed at 2min and the demold criterion is 60% conversion everywhere throughout the part.

Three factors--mixing, degradation temperature and minimum conversion for demolding at 2 min--then determine the limits on initial material and mold temperatures for molding good parts. Similar diagrams could be constructed for polyurethanes and polyureas if minimum t_f and t_d can be specified.

Moldability diagrams can be very helpful for initial studies on a new chemical formulation or estimating productivity on a new mold. However, they do not tell us much about what conditions give optimum final properties. The influence of the process on property development will be discussed further in the last section of this chapter, but first we will look at some aspects of mold design and the final steps in the RIM process, part finishing.

6.5   Mold Design

RIM molds are not fundamentally different from those used in thermoplastic, especially elastomer, injection molding. For high volume production RIM requires machined steel molds, just like TIM. For lower volume cast metals and even epoxies can be used because of the lower pressures. Because of the low viscosities during filling RIM molds must seal tighter around cores, pins and mold edges. Venting is also much more critical in RIM than in TIM, as is the need to remove heat due to the reaction exotherm.

There are a number of good discussions of RIM mold design available (Walner, 1978; Becker, 1979; Misitano, 1979; Mobay, 1982; Stocki, 1983; Oertel, 1985; Sweeney, 1987). Here we will review the key points. In Chapter 5 design criteria for aftermixers, runners and gates were presented. The key issue in mold filling is to avoid air entrapment and large bubbles.

As shown in Figure 5.25 and 26 venting out air is critical. Presses are often tiltable to help locate vents at the highest points. Gates should be mounted at the lowest part of the mold. The appearance surface of the part should be on the bottom of the mold. Large bubbles are more likely to collect on the top. To prevent liquid from flashing out around all the edges a land (6 mm wide for steel, 13 mm for aluminum) is machined around the perimeter of the parting line. The area outside this sealing edge should be relieved approximately 1mm. Sometimes gaskets are used to seal molds. The gasket can even be created by allowing the RIM material itself to flow into a dovetailed groove (Mobay, 1982). Vents can be cut through the seal at the points where air needs to escape. A channel 1.5 x 3 mm will adequately vent the mold and be strong enough to stay with the part on demolding. If flash stays on the mold it can interrupt sealing of the next part or fall into the mold cavity leading to a defect. Flash is still a significant problem for RIM. Many TIM parts are molded without any flash.

For lower volume production RIM molds can be self contained, clamped up by hand or with hydraulic ejectors built in around the sides of the mold (Misitano, 1979). But most molds are mounted in hydraulic presses. Because RIM molds are under low pressure (recall Figure 6.9) presses require much less force than for TIM. This is one of the main economic advantages of RIM over TIM with large parts. However, demolding can require larger pressures, up to 10 bar, especially if a part does not fully cure due to poor mixing or a formulation error. With large parts this leads to large forces. A 1 m x 1 m flat part can require up to 100 tons force. This force dictates the size of hydraulic pumps and platen stiffness.

To aid demolding, air ejectors or knock out pins can be used. Knock out pins are located on the top of the mold, the non-appearance side, and must be large enough in diameter to avoid tearing the part. Knock out pins must be liquid but not air tight. This is also true for moving cores. Clearances less than 0.05 mm are recommended (Oertel, 1985, Chapter 7). Robot arms with suction cups are also being evaluated to automate demolding.

There are several mold materials and construction methods to choose from. Basis for choice include surface quality, number of parts, dimensional tolerance and cost. Table 6.1 compares the important mold materials. The best surfaces, highest production and tolerances are obtained with machined metal. Steel is about 15% more expensive but has about double the life time and better surface than aluminum. Steel can be nickel treated to further improve surface quality and demolding. It is used for most automotive RIM molds.

Cast metal molds are significantly cheaper but suffer reduced dimensional control. Kirksite, a zinc alloy, is preferred among the cast materials. Stainless steel tubes can be cast directly into Kirksite, an economical way to install cooling lines.

High surface quality with excellent dimensional control can be achieved by electrolytically or vapor deposited nickel shells. Cooling lines are attached to the back side of the shell which is embedded in epoxy and mounted on a steel frame for structural rigidity. Such shells are thin which makes them difficult to repair or modify.

Cast or hand lay up epoxy molds are the cheapest but are only suitable for prototyping or limited runs of simple shapes. Flexible silicone rubber molds can also be used for prototypes.

Temperature control can be more important in RIM than TIM because it controls the chemical reaction rate at the mold surface. Temperatures should be maintained to ±2°C. For most RIM materials polyurethanes, ureas and dicyclopentadiene the range is 50-80°C which can be controlled by water circulation. Nylon, epoxy and acrylamate require 100-140° (see

Table 6.1 Comparison of RIM Mold Materials and Methods of Construction[a]

CONSTRUCTION METHOD:	MACHINED		CAST			SHELL		LAY UP
Material:	steel	aluminum	aluminum	kirksite	plastic	nickel	sprayed metal	epoxy
Surface Quality	E	VG	P	G	F	E	G	F
Durability (# of parts)[b]	10^6	5×10^5	5×10^5	5×10^5	5000 (prototype)	75,000	15,000	5000 (prototype)
Temperature Control	VG	E	E	VG	P	F	F	P
Mold Repair	E	E	G	VG	G	P	P	F
Engineering Change Incorporation	E	E	F	G	F	P	P	F
Dimensional Reproduction from Models	VG	VG	P	F	F	E	VG	G
Approximate % Cost[c]	100%	85%	56%	43%	26%	38%	29%	21%
Mold Delivery (weeks)[3]	20-22	18-24	16-18	16-18	6-8	14-16	14	11

E = Excellent,   VG = Very Good,   G = Good,   F = Fair,   P = Poor

(a) Stocki (1983)
(b) Based on a simple, easy to mold part. Reduce by 10 times or more for difficult parts.
(c) Mold size approximately 145cm long x 76 cm wide x 69 cm shut height. 100% cost ≈ $90,000 in 1983 dollars.

Table 1.2 and Chapter 7) which necessitates oil tempering. For the first few shots in a mold heating is required but soon reaction heat builds up and at steady state cooling is required. Heat removal rates up to 450 J/gm are required. In thick sections removal rate can be increased by higher velocity using smaller cooling tube diameter (Stocki, 1983). Stocki recommends placing water lines 5-8 cm apart, 2 cm below the surface and 9-13 mm diameter.

6.6 Finishing

After demolding, as indicated in Figure 1.2, there are usually several finishing steps before a RIM part is ready to deliver. Table 6.2 summarizes these operations in the order they are usually performed.

Table 6.2   Finishing Operations

- trim flash, gate

- inspection

- repair large bubbles, pin holes

- postcure

- wash off external mold release

- prime, paint

On large parts trimming is often done manually by the operator between shots. Flash and the aftermixer, runner, gate combination are cut off with a knife. A sander is sometimes used to smooth off the cut areas. Interior trimming and holes are opened with a router. Elimination of flash on RIM parts would reduce labor requirements for the process. For smaller parts deflashing can be automated by die cutting or a programmed sander or grinder. Recessing parting lines can sometimes facilitate flash removal. All the flash will be on one side making it easier to remove with a rotary wire brush (Sweeney, 1987, pg 246). In automotive fascia (bumper covers) 4% of polymer is removed by trimming. This currently goes to land fill. There is a major opportunity for recycling perhaps as filler or reverse reaction to regenerate some of the starting components.

Preliminary inspection is also done by the operator who looks for tears, large bubbles, imbedded flash and porosity. A second inspection is often done on large, complex parts. A light table aids inspection of non-pigmented parts. If the part is to be painted many of these defects can be repaired by drilling out the area and filling it with a urethane paste. Scrap rates are 2-4% for automotive fascia production in the US (Cornell, 1988).

Large parts are placed on aluminum frames to prevent warpage as they cool down and during postcuring. These frames also serve as supports and conveyors through washing and painting lines. Typically postcure for exterior automotive parts is 120°C for one hour. The purpose of postcuring is mainly to drive out CO_2 and perhaps some of the internal mold release to prevent problems in painting. The improvement in properties especially for polyurethane urea systems is not significant. Modular automotive window frames, which receive a coating in the mold, and other non painted parts are not postcured.

Before painting mold release and dust must be washed from the parts. This is done with phosphoric acid and rinsed with hot water. This cycle is repeated followed by a final rinse with deionized water at 90-95°C.

Painting for automotive applications is done "off line". First a primer is applied and baked at 120°C for one hour followed by one or more polyurethane top coats which are baked at the same temperature for shorter times. There is interest to be able to paint plastic parts along with metal parts all assembled on the car body. Currently "on line" painting requires 175°C for priming and 200°C for top coats. These temperatures, especially for large parts, push the dimensional stability limits of most polymers. Nylon-6 and some polyureas are claimed to be suitable for on line painting (Martinez and Vanderhider, 1987).

6.7   Influence of Processing on Properties

Figures 5.19, 6.24 and 6.25 define moldability windows for crosslinking and phase separating systems. These windows mainly help to determine whether a part can be made at all by RIM. Within these windows further property optimization is possible. In this section we will review the influence of each process step on final part properties.

Proper metering of reactants is critical to control molecular weight as indicated in Figure 2.3. Fruzzetti et al. (1977) showed the great sensitivity of flex life to imbalance in reactant ratios.

As we saw in Figure 6.10 increased gas loading in the reactant tanks reduces part density. This generally reduces modulus and impact strength but it will of course reduce cost and also help control mold shrinkage (Ostfeld, 1982).

Sufficient mixing is also critical for high molecular weight. At low Reynolds number Kolodziej et al. (1986) report better phase separation but lower molecular weight (Figure 4.12) and the presence of hard segment globules (see Figure 2.25). Nishimura et al. (1986) also found reduced phase separation at high Reynolds numbers, by electron microscopy, x ray and sensitivity of the modulus to temperature. Flex life appears to be sensitive to increasing Re above the normal critical value (Table 4.1; Fruzzetti et al., 1977). Incomplete mixing may also be part of the cause of brittleness problems for polyureas (Willkomm, 1988).

Flow rate into RIM molds has little effect on orientation and thus mechanical properties are fairly isotropic in contrast to TIM (Castro et al., 1984). This is not true, though, for fiber filled RIM, as will be discussed in Chapter 8. Weld or knit lines are also not a problem except near the gel point (Figure 5.28). High filling velocities can lead to air entrainment and problems with large bubbles as discussed in Chapter 5.

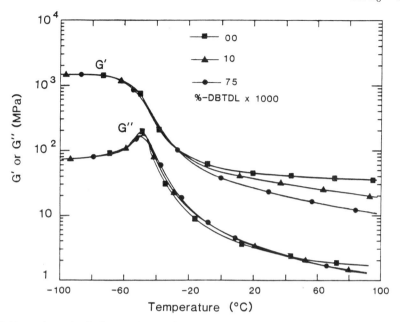

Figure 6.27 Dynamic mechanical properties as a function of catalyst concentration for a linear, ethylene glycol based polyurethane similar to F-2 in Table 2.4 (from Camargo et al., 1985).

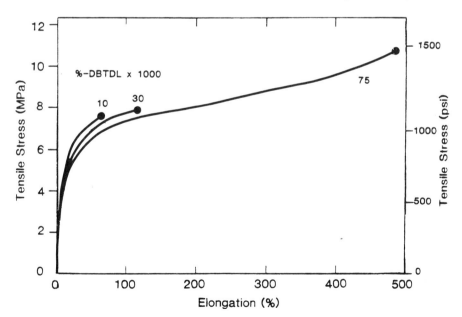

Figure 6.28 Stress-strain curves as a function of catalyst concentration for same formulation as in Figure 6.27 (from Camargo et al., 1985).

Table 6.3 Typical Properties of Polyurethane and Polyurea RIM Polymers, Unreinforced

	Method[e]	BDO based[a]	urea urethane[b]	urea[c]	EG based[d]
E_f, Flex. Mod. MPa	D790	160	365	540	830
$\dfrac{E_f (-29°C)}{E_f (70°C)}$		6	3.4	3.0	~5
Tensile Strength MPa	D638	20	28	24	28
Elongation, %	D638	200+	280	200+	120
Gardner Impact J	D2794	8-10	--	--	1.6-2
Izod Impact J/m	D256	(high)	480	590	240
Heat Distortion temperature, °C	D648	60	~135	--	80
Heat Sag at 121°C mm		10	4	16	10
Thermal Expansion m/m°C x 10[6]	D696	155	160	180[c]	150
Specific Gravity		1.00	1.04	1.06	1.00

a) Union Carbide's RIM 2200, F-3 in Table 24. Lee (1980).
b) Similar to F-5 in Tabel 2.4 Mobay's Bayflx 110-50, Dow's Spectrim 50.
c) Similar to F-7 Table 2.4. See also Table 8.2. Heat sag at 135°C.
d) Union Carbide's RIM 125, Lee (1980). Bayer Bayflex 103 similar. Higher modulus obtainable with highly crosslinked urethanes like Mobay's Baydur 726.
e) See Table 7.1 for test descriptions.

In the curing stage mold temperature is the key variable. Typically increased T_w means more part shrinkage (Ostfield, 1982). For phase separating polyurethane systems we saw in Figure 2.28b that at high temperatures, $T > 120°C$, modulus decreases. This is because molecular weight builds quickly and interferes with phase separation. Tensile strength is not affected as much. Even if mold temperature is below 120°, due to reaction exotherm and speed, the center of a part will get much hotter. Thus thicker parts may have a lower modulus (Castro et al., 1984) and can have severe morphology gradients (Chang et al., 1982). At low T_w parts become brittle with poor surface quality as shown in Figures 6.19-6.22.

For polyureas mold temperature has a large influence on brittleness (Figure 6.23). High mold temperature or long times in the mold appear to promote enough buildup in molecular weight to yield a tough polymer.

Postcuring also influences final properties. Simply aging polyureas at room temperature improves their toughness. Typical postcuring is at 120°C for about 1 hr. As mentioned in the previous section, the paint oven can add to this time. Dominguez (1981) reported that this postcuring of an ethylene glycol based polyurethane raised its room temperature modulus by 15% and reduced heat sag at 165°C about 30%. However, 1 hr at 165°C reduced heat sag by a factor of 10.

The other major influence on final properties is of course the formulation. The role of crosslink density, hard segment type, concentration and length were discussed in Chapter 2.7. It is possible to make some formulation changes in a production environment. Catalyst level is one which can have a strong influence on final properties as shown in Figures 6.27 and 6.28. High catalyst level acts like higher mold temperature (compare Figures 2.24a and b); more catalyst speeds up reaction kinetics with respect to phase separation kinetics. This leads to proper phase separation, a more temperature dependent modulus in Figure 6.27, but higher molecular weight and thus higher elongation as shown in Figure 6.28.

When all the process steps are properly combined, parts with excellent and very reproducible properties are molded routinely. Table 6.3 gives properties for several of the polyurethane and polyurea formulations given in Table 2.4.

REFERENCES

Alberino, L.M.; Regelman, D.F.; Vespoli, N.P. "Process for preparation of polyurea elastomers," US Patent 4,546,114 (1985).

Aronhime, M.T.; Gillham, J.K. "Time-temperature-transformation (TTT) cure diagram of thermosetting polymeric systems," *Adv. Polym. Sci.*, **1986**, *78*, 83-113.

Barone, M.R.; Caulk, D.A.; "Nonuniform cavity surface temperatures in injection molding," *Polym. Proc. Eng.* **1985**, *3*, 149-158.

Batch, G. PhD Thesis, Department of Chemical Engineering and Materials Science, University of Minnesota, 1989.

Becker, W.E., ed. *Reaction Injection Molding*, Van Nostrand Reinhold: New York, 1979.

Begemann, M.; Maier, U.; Müller, H.; Pierkes, L. "RIM-improved technology of the machines and selective mould design to enhance moulding quality," translated from the Conference Handbook of the 13th Inst. für Kunststoffverarbeitung Kolloquium, Aachen, Block 13, p 405 (1986).

Blake, J.W. "Studies in reaction injection mold filling," PhD Thesis, University of Minnesota, 1987.

Blake, J.W.; Macosko, C.W.; "Foaming in microcelleular RIM," *J. Cellular Plast.* **1988**, submitted.

Broyer, E.; Macosko, C.W.; "Heat transfer and curing in polymer reaction molding," *AIChE J.* **1976**, *22*, 268-276.

Broyer, E.; Macosko, C.W.; Critchfield, F.E.; Lawler, L.F.; "Curing and heat transfer in polyurethane reaction molding," *Polym. Eng. Sci.* **1978**, *18*, 382-386.

Camargo, R.E., Macosko, C.W., Tirrell, M. and Wellinghoff, S.T. "Phase separation studies in RIM polyurethanes: catalyst and hard segment crystallinity effects," *Polym.* **1985**, *26*, 1145-1154.

Castro, J.M.; Macosko, C.W. "Studies of mold filling and curing in the reaction injection molding process," *AIChE J.* **1982**, *28*, 250-260.

Castro, J.M.; Macosko, C.W.; Critchfield, F.E.; "Effect of processing conditions on premature gelling, knit line strength, and physical properties for the RIM process," *J. Appl. Polym. Sci.* **1984**, *29*, 1959-1969.

Cekoric, M.E.; Taylor, R.P.; Barrickman, C.E.; "Internal mold release, the next step forward in RIM," SAE Tech. Paper #830488, Detroit, March 1983.

Chang, A.L.; Briber, R.M.; Thomas, E.L.; Zdrahala, R.J.; Critchfield, F.E. "Morphology study of the structure developed during the polymerization of a series of segmented polyurethanes," *Polym.* **1982**, *23*, 1060-1065.

Cornell, M.C. Dow Chemical, personal communication 1988.

Dominguez, R.J.G. "The effect of annealing on the thermal properties of reaction injection molded urethane elastomers," *Polym. Eng. Sci.* **1981**, *21*, 1210-1217.

Dominguez, R.J.G.; Rice, D.M.; Grigsby, R.A. "Polyurea heralds latest generation of RIM systems," *Plast. Eng.* **1987**, *43 no. 11*, 41-44.

Estevez, S.R.; Castro, J.M.; "Applications of a reaction injection molding process model in the analysis of premature gelling, demold time, and maximum temperature rise," *Polym. Eng. Sci.* **1984**, *24*, 428-434.

Fruzzetti, R.E.; Hogan, J.M.; Murray, F.J.; White, J.R. "Factors affecting the quality of impingement mixed RIM urethanes," Soc. Auto. Eng., Passenger Car Meeting, Detroit, Sept. 1977.

Gent, A.N.; Tobias, R.H. "Threshold tear strength of some molecular networks," *ACS Symp. Ser.* **1982**, *193*, 367-376.

Lee, J.L.; Macosko, C.W.; "Heat transfer in polymer reaction molding," *Int. J. Heat Mass Transfer* **1980**, *23*, 1479-1492.

Kamal, M.R.; Singh, P.; Samak, Q.; Kakarala, S.M.; "Microstructure and mechanical behavior of reinforced reaction injection molded (RRIM) polyurethane," *Polym. Eng. Sci.* **1987**, *27*, 1258-64.

Kemp, D. Dow Corning Corp., personal communication 1985.

Kim, D.H.; Kim, S.C. "Engineering analysis of reaction injection molding process of epoxy resin," *Polym. Composites*, **1987**, *8*, 208-217.

Kolodziej, P., Yang, W.P., Macosko, C.W. and Wellinghoff, S.T. "Impingement mixing and its effect on the microstructure of RIM polyurethanes," *J. Polym. Sci., Phys.*, **1986**, *24*, 2359-2377.

Martinez, E.C.; Vanderhider, A. "On line paintable RIM body panels," Soc. Auto Eng., Int. Cong. and Exp., Detroit, Feb. 1987.

Macosko, C.W.; Lee, L.J.; "Heat transfer and property development in liquid silicone rubber molding," *Rubber Chem. Tech.* **1984**, *58*, 436-448.

Manas-Zloczower, I.; Macosko, C.W.; "Moldability diagrams for reaction injection molding," *Polym. Process Eng.*, **1986**, *4* , 173-184.

Manas-Zloczower, I.; Macosko, C.W.; "Moldability diagrams for reaction injection molding of a polyurethane crosslinking system," *Polym. Eng. Sci.* **1988**, *28*, 000.

Manzione, L.T.; Osinski, J.S.; "Moldability studies in reactive polymer processing," *Polym. Eng. Sci.* **1983**, *23*, 576-585.

Meyer, L.W. "Self-releasing urethane molding systems: productivity study," in "Reaction Injection Molding," Kresta, J.E., Ed., Am. Chem. Soc. Symp. Series 270, Washington, D.C., 1985.

Misitano, G. "Guidelines for the effective design of RIM molds," *Plast. Eng.* **1979**, *35 no 2*, 27-31.

Mobay, "Bayflex polyurethane elastomer, mold design manual," Mobay Chemical, Polyurethane Division, Pittsburgh, 1982.

Oertel, G., ed. "Polyurethane Handbook," Hanser: Münich, 1985.

Ostfeld, H.G. "Polyurethanes in automotive: technical aspects for their use in passenger vehicles," Soc. Plast. Ind., Urethane Tech./Market Conf. 1982, 149-159.

Okuda, K. "Computer analysis of RIM moldability from the density distribution in molded products," in *Reaction Injection Molding and Fast Polymerization Reactions*, ed. J.E. Kresta, Polymer Science and Technology, vol. 18, Plenum Press, 279-298, 1982.

Perry, S.J.; Castro, J.M.; Macosko, C.W.; "A viscometer for fast polymerizing systems," *J. Rheol.* **1985**, *29*, 19-36.

Plevyak, J.E.; Sobieski, L.A. "Improved RIM processing with silicone internal mold release technology," in *Reaction Injection Molding, Polymer Chemistry and Engineering*, J.E. Kresta, ed. ACS Symp. Ser. 270, 1985, 213-224.

Sweeney, F.M. "Reaction Injection Molding: Machinery and Processes," Marcel Dekker: New York, **1987**.

Stocki, W.; "Selecting mold-making materials depends on production, part size," *Rubber and Plast. News* **1983**, *October 10*, 34-37.

Tirrell, M.; Lee, L.J; Macosko, C.W.; "Conversion and composition profiles in polyurethane reaction molding," ACS Symposium Series, 104, J.N. Henderson and T.C. Bouton, eds., Amer. Chem. Soc., Washington, D.C., **1979**, 149.

Vespoli, N.P. Personal communication, 1985.

Vespoli, N.P.; Alberino, L.M.; "Computer modeling of the heat transfer processes and reaction kinetics of urethane-modified isocyanurate RIM systems," *Polym. Proc. Eng.* **1985**, *3*, 127-147.

Wallner, J.; "RIM: Tips on tooling," *Plast. Tech.* **1978**, *24* No. 3, 61-64.

Willey, S.J.; Nied, H.A.; "Scaling concepts in thermoset injection molding," Proceedings of Second International Conference on Reactive Polymer Processing, Pittsburgh, Oct. 1983.

Williams, R.J.J.; Rojas, A.J.; Marciano, J.H.; Ruzzo, M.M.; "General trends in the curing of thermosets in heated molds," *Polym. Plast. Technol. Eng.* **1985**, *24*, 243-266.

Willkomm, W.R.; Chen, Z.S.; Macosko, C.W. "Properties and phase separation of RIM and solution polymerized polyureas as a function of hard block content," *Polym. Eng. Sci.* **1987**, *27*, 000.

Willkomm, W.R.; Jennings, R.; Macosko, C.W.; "Methods for evaluating mold releases in reaction injection molding," in preparation 1989.

Yang, W.P. "Phase separation dynamics in polyurethane reaction injeciton molding," PhD Thesis, University of Minnesota 1987.

Yang, W.P.; Macosko, C.W. "Phase separation during fast (RIM) polyurethane polymerization," *Makromolekular Chemie*, **1989**, to appear.

7

NON-URETHANES FOR RIM

Because RIM has developed around polyurethane and polyurea chemistry, this book has concentrated on them. Problems in material delivery, mixing and molding have all been attacked using urethanes. However, RIM machines developed for polyurethanes can readily be adapted to a number of other polymerizations. Futhermore, several new chemical systems have been developed specifically for RIM. It is useful to study these non-urethane RIM systems for two reasons. First, they are becoming important commercially, and second, by understanding what chemistry does and does not work for reaction injection molding, we will have a much better understanding of the process itself. Thus in several ways this chapter is also a review of the important points from Chapters 3 through 6.

Looking back to Chapter 1 in Table 1.2 we find a summary of the chemical systems used today in RIM. Each is discussed in more detail in the following sections; however, some of the motivation for introduction of these non-urethanes can be appreciated by looking at Table 7.1. There is considerable interest in developing a RIM polymer with high modulus and toughness, an engineering thermoplastic for automotive body panels and other structural applications. As pointed out in Chapter 2.7, when urethanes are formulated with high hard segment content to give a flexural modulus over 1000 MPa they become brittle. The polyurethane in Table 7.1 represents about the highest possible modulus still having good impact strength. By going to a crystalline polyamide like nylon 6 or to highly crosslinked polymers, like polydicyclopentadiene (DCP) or epoxies, much higher moduli are possible and in some cases higher use temperatures as well. We can see that, although their modulus is high, epoxies also suffer from low impact strength. They are only useful in RIM when reinforced with continuous fibers. Properties and special process considerations for reinforced RIM polymers will be discussed in the next Chapter.

How can we determine what polymerizations will work for RIM? What requirements does the RIM process put on the chemical system? To answer these questions it is helpful to review each of the eight RIM unit operations in Figure 1.2 to see what limitations each imposes on the chemistry. These are summarized in Table 7.2.

Table 7.1 Typical Properties of RIM Polymers, Unreinforced

property	method[e]	high mod. PU[a]	Nylon [b]	DCP [c]	Epoxy [d]
E_f MPa	D790	830	1400	1860	2900
$\dfrac{E_f(-29°C)}{E_f(70°C)}$		~5	4.2	1.6	1.7
Tensile Strength MPa	D638	28	44	34	69
Elongation, %	D638	120	200	70	6
Gardner Impact J	D2794	1.6-2	18	15 (c)	4
Izod impact J/m	D256	240	300	430	24
Heat Distortion Temperature, °C	D648	80	175	95	118
Heat Sag[e] mm	121°C 163°	10	-- 2.5	61 (c)	1 9
Thermal Expansion m/m°C x10^6	D696	150	122	54	56
Specific Gravity		1.00	1.13	1.04	

(a) Union Carbide RIM 125; Bayer Bayflex 103; Lee (1980)

(b) Nyrim P2000; Hedrick, Gabbert and Wohl (1985); Gabbert, Garner and Hedrick (1983).

(c) Hercules' Metton product literature; Geer and Stoutland (1985); plate rather than Gardner impact; heat distortion at 135°C, 1 hour, 152 mm overhang, Geer (1983).

(d) de la Mare and Brownscombe (1983)

(e) ASTM test numbers; heat sag is a General Motors test method, deflection of the end of a bar 3.2 mm thick overhung 100 mm for 1 hour at 121°C or 30 min at 163°C.

Supply - Ideally reactants should be liquids at room temperature and stable for weeks in a sealed supply system like that shown in Figure 3.6. Tank material is typically steel. Many of these storage tanks can be heated up to 60 or 70°C, but they often have cold spots where reactants could freeze. The reactants must be pumped to blend tanks and to the conditioning tanks on the machine. Highly filled liquids require special screw pumps. Solid monomers like caprolactam for nylon 6 can sometimes be fed with a sealed augur to the conditioning tanks where they melt.

Condition - Nearly all of the conditioning systems on RIM machines in use today are limited to temperatures below 70°C. Furthermore, because they have uninsulated sections they require materials which remain liquid at ambient temperature. A few machines are completely heat traced and can process monomers like caprolactam or epoxies which are solid at room temperature. Besides temperature control, dry air or an inert gas is often dispersed in at least one of the components for later mold packing. Typically this requires addition of a surfactant. Under pressure this gas must dissolve and then come out of solution rapidly during the curing step. Dry nitrogen is also used to protect monomers from water vapor and oxygen.

Meter - Nearly all RIM machines, like the one shown in Figure 3.1, are designed for two components. Addition of a third component stream is possible particularly if it is of low flow rate. A small multistroke pump can be used since most mixheads have 4 entry ports. Since RIM machines are designed for the tight control needed for the urethane step addition reaction, such machines should provide adequate ratio control for other polymerizations. For example nylon 6 and DCP RIM formulations are chain reactions and tolerate ±5% imbalance.

Mix - As discussed extensively in Chapter 4 good impingement mixing requires a critical Reynolds number of about 300. For a given machine this leads to a maximum viscosity at the mixing temperature. Typical small production RIM machines have a maximum flow rate of 150 g/s per stream. Assuming a minimum mixhead orifice of 1mm diameter and a specific gravity of 1 gives a maximum viscosity for good mixing of 0.6 Pa·s. The largest RIM machines have about 2 kg/s flow rate and thus could handle 10 times this maximum viscosity. The microscopic observations of micromixing implies that there may also be some level of compatibility required between the two reactants to allow mixing.

Fill - Fast reaction rates are desirable for fast demolding but there must be some delay to allow time for mixing and for mold filling. The later is generally more restrictive. As shown in Chapter 5 the time for the reaction to gel under adiabatic conditions, $t_{g,ad}$, must be less than the time to fill the mold. With large parts and current equipment the gel time must be greater than 1s.

High filling velocities can lead to gas entrainment and to large bubbles in the part. This seems also to be controlled by the Reynolds number, now based on mold thickness and velocity in the mold cavity. This Re restriction implies a minimum on viscosity for bubble free filling. For example, dicyclopentadiene monomer has a viscosity of 3 mPa·s. To use it for RIM formulation some elastomer is added to raise the viscosity and achieve smoother filling.

Curing - Again because current equipment was developed for urethanes, curing is carried out at low mold temperatures. It is relatively simple to have hotter molds, much simpler than heat tracing the conditioning and metering side of the machine. However, some of the advantage of RIM is lost with hot molds since they are harder to work with, more expensive and, of course, have higher energy costs. To prevent polymer degradation, mold temperature plus the reaction exotherm must not excede the degradation temperature.

Some polymerizations generate a small molecule by-product. The reaction of phenol with formaldehyde to produce phenolic resin, for example, generates water which can vaporize at the reaction temperatures. This might be used to expand the polymer to compensate for shrinkage but excessive blowing will lead to poor properties.

Demolding - As Figures 6.24 and 6.25 in Chapter 6 indicate, demolding also puts restrictions on mold wall temperature. If T_w is too near the melting or phase mixing temperature solidification will be slow and delay demolding or the part may warp after demolding. If T_w is too low one of the monomers or reaction products may precipitate prematurely or the mixture may vitrify before the polymer is strong enough to demold.

Demold time should be less than 3 min. (less than 45 s for high volume production). At demolding the reaction should typically be at least 95% complete and solidified (due to crosslinking or phase separation). The part should release from the mold without external agents applied to the mold surface except perhaps at the beginning of a shift.

Table 7.2   RIM Material Requirements

Supply	• stable $\geq$ 1 week
	• pumpable
Condition	• $T_o < 60°C$
	$< 150°C$ (high temperature machine)
	• gas dispersion
Meter	• two components
	• $\pm 0.5\%$ stoichiometry
Mix	• $\eta < 1$ Pa·s
	• compatibility
	• $t_{g,ad} > 0.1s$
Fill	• $t_{g,ad} > t_f > 1s$
	• $\eta > 10 - 100$ m Pa·s
	to prevent bubbles
Cure	• $T_w < 100°C$
	$< 200°C$ (high temperature mold)
	• $T_w < T_{degrad} - \Delta T_{ad}$
	• control by-products
	• compensate for shrinkage
Demold	• $T_w <$ melting or phase mix temperature
	• $T_w >$ glass transition, precipitation
	• $t_d < 3$ min (sufficient green strength)
	< 45 s (high production)
	• easy mold release
Finish	• little flash
	• minimize post cure
	• paintable

Finish - Ideally there should be no flash on the part. This is difficult to achieve with the low viscosity liquids used in RIM. What flash there is should come off the mold surface cleanly staying with the piece. Ideally, postcuring should not be necessary, but sometimes it is available without cost as part of a painting operation. Postcuring often seems necessary to drive out any gas and complete the last part of the reaction. Mold surfaces typically must be cleaned carefully for painting.

Before we go on to see how some polymerization systems have been sucessfully designed for RIM, it may be helpful to consider some which should *not* work. Not all fast polymerizations are suitable for RIM. Obviously, polymerizations in solution, emulsion or suspension are not suited for direct production of a molded part. Some polymers, such as polyethylene or polypropylene, are presently only made in solution or dispersion.

Solvents or slow reaction rates are often used to prevent high temperatures and thermal degradation which can result in very exothermic reactions like the polymerization of styrene, methyl methacrylate and other vinyl polymerizations. To prevent huge exotherms most RIM systems incorporate some prepolymer or oligomers mixed with monomers.

Polymers such as polyesters which require extensive condensation of a small molecule are also poor RIM candidates. Monomers or oligomers which must be polymerized near or above the melting temperature of the polymer won't work; it is impractical to first heat the mold for reaction then cool it to cause crystallization. The role of micromixing in impingement mixing is not well understood, but it seems for example that some soft segment oligomers are "too incompatible" to produce suitable RIM polyurethanes. And clearly high initial viscosity also won't work.

7.1 Polyamide 6

Like polyurethane, nylon 6 evolved into RIM from casting technology. To date it is the most studied non-urethane system. Its advantages of high modulus with high impact strength and high temperature stability are clear from Table 7.1.

Anionic polymerization of ε-caprolactam has been used for many years to make cast nylon 6. With an acyl lactam or similar initiator and a metal catalyst it procedes rapidly at 100-160°C, well below the melting temperature of nylon 6, 220°C. Figure 7.1 shows the two step propagation mechanism. This propagation is initiated by introducing premade lactam chain ends.

The metal catalyst forms an anion which rapidly opens the ring at the chain end and replaces it with another lactam. The cycle is completed when the N^- in the chain abstracts

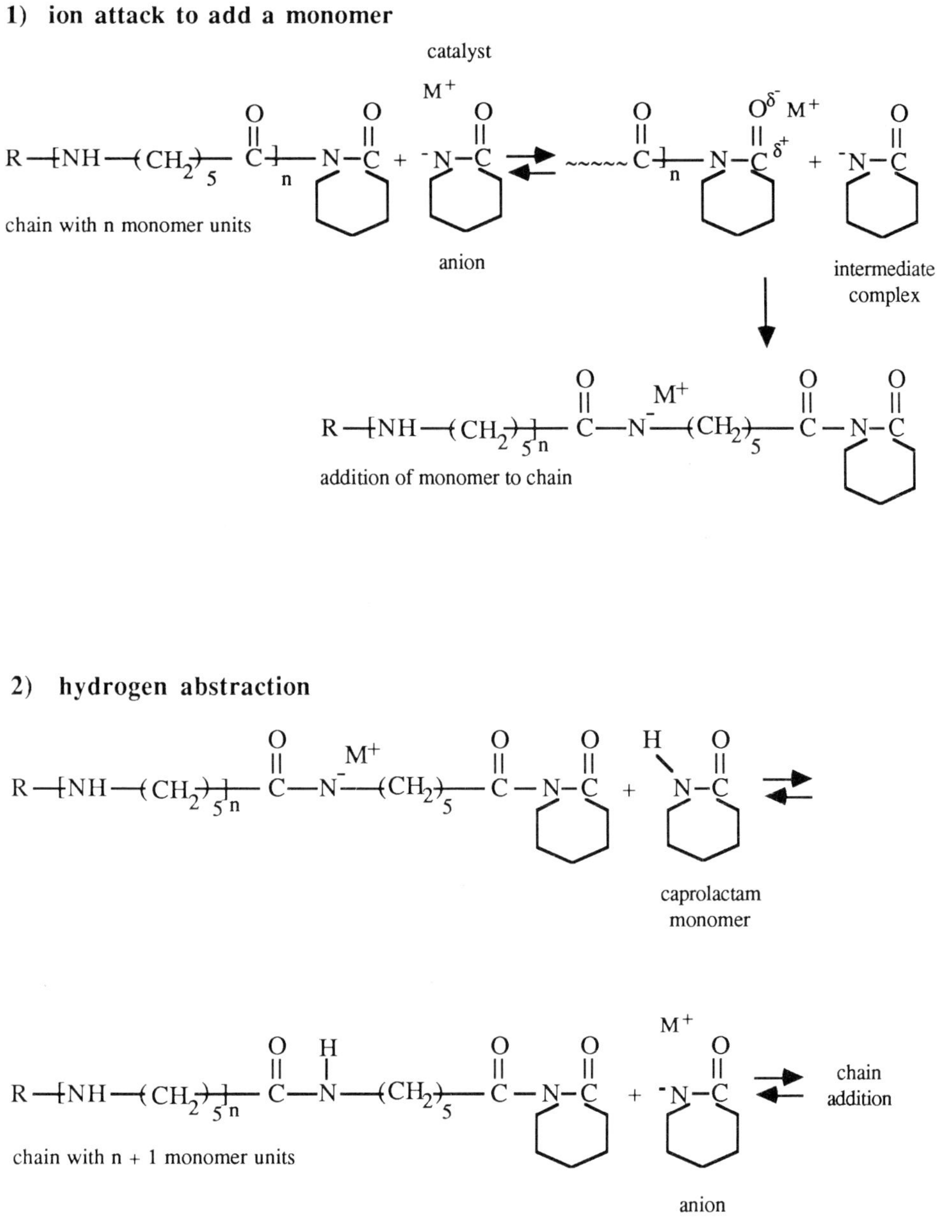

1) ion attack to add a monomer

catalyst

M^+

R—(NH—(CH$_2$)$_5$—C)$_n$—N–C + $^-$N–C $\rightleftharpoons$ ~~~~C)$_n$—N–C$^{\delta+}$ + $^-$N–C

chain with n monomer units

anion

O$^{\delta-}$ M$^+$

intermediate complex

R—(NH—(CH$_2$)$_5$)$_n$—C—N$^-$—(CH$_2$)$_5$—C—N–C

M$^+$

addition of monomer to chain

2) hydrogen abstraction

R—(NH—(CH$_2$)$_5$)$_n$—C—N$^-$—(CH$_2$)$_5$—C—N–C + N–C $\rightleftharpoons$

M$^+$

H

caprolactam monomer

R—(NH—(CH$_2$)$_5$)$_n$—C—N—(CH$_2$)$_5$—C—N–C + $^-$N–C $\rightleftharpoons$

H

M$^+$

chain addition

chain with n + 1 monomer units

anion

Figure 7.1   Two step propagation mechanism for the anionic polymerization of ε-caprolactam to polyamide 6. The catalyst M$^+$ is typically MgBr$^+$ or Na$^+$.

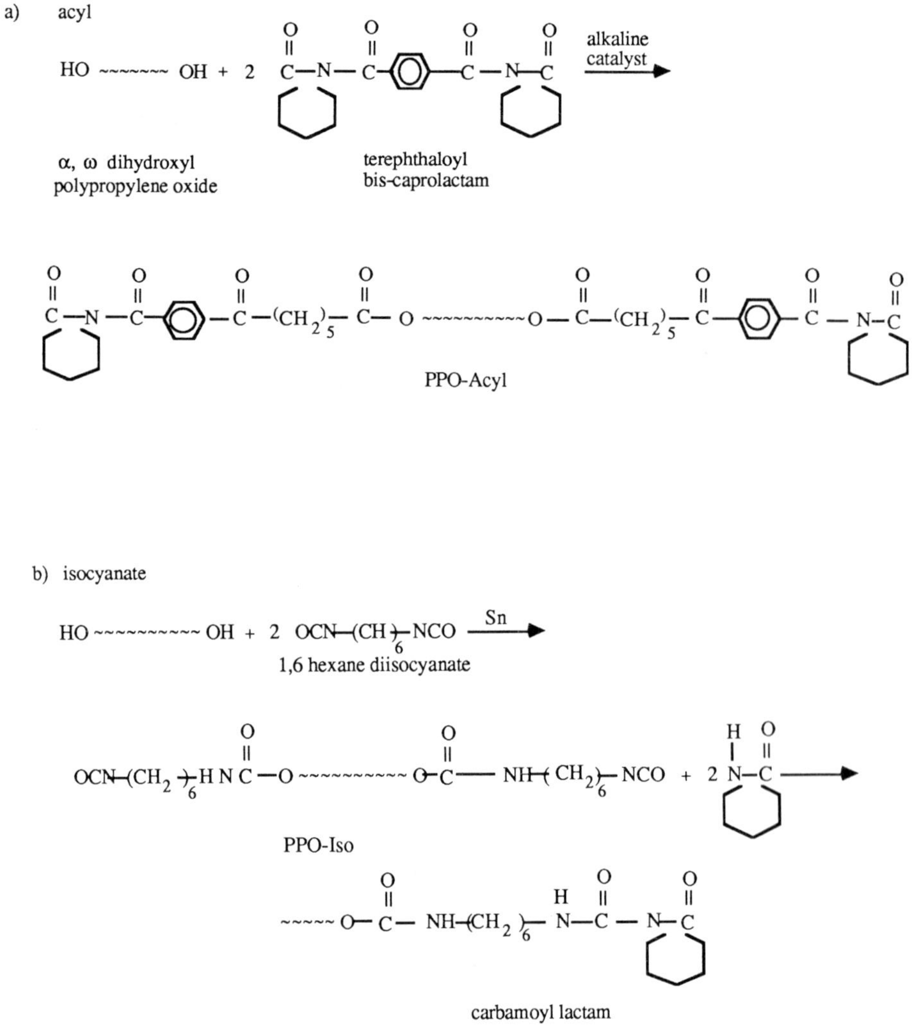

a)   acyl

α, ω dihydroxyl
polypropylene oxide

terephthaloyl
bis-caprolactam

alkaline
catalyst

PPO-Acyl

b) isocyanate

HO ~~~~~~~~~ OH + 2 OCN—(CH$_2$)$_6$—NCO $\xrightarrow{\text{Sn}}$

1,6 hexane diisocyanate

PPO-Iso

carbamoyl lactam

Figure 7.2   Two methods for forming acyl lactam initiator groups on the ends of an OH terminated oligomer a) Hedrick, Gabbert and Wohl (1985) b) Sibal, Camargo and Macosko (1984) van der
Loos and van Geenan (1985).

a hydrogen to form another active lactam cation. The most commonly used catalysts are sodium caprolactam and magnesium bromide caprolactam (Hedrick, Gabbert and Wohl, 1985).

Impact strength of the pure homopolymer can be improved considerably by introducing elastomeric blocks. This is done easily by putting initiator groups on the ends of an elastomeric oligomer. The two most common types of initiator are illustrated in Figure 7.2. Blocks of nylon 6 then form on each end of the oligomer leading to an ABA triblock structure. However, near the end of the reaction as monomer is depleted the hydrogen abstraction step can lead to branching. It is also possible that the ester linkages to the oligomer (Figure 7.2a) can serve as weak initiation or transfer sites. Furthermore, at high temperature the amide groups may interchange. All of these side reactions lead to more of an alternating block copolymer structure and to a lightly crosslinked network as shown schematically in Figure 7.3 (Kurz, 1985).

Like polyurethanes, the soft blocks segregate into fine lamellar domains. Van der Loos and van Geenen (1984, 1985) have demonstrated the dramatic difference between simply blending in polypropylene oxide and incorporating it as blocks in the chain. Table 7.3 shows the large increase in impact strength and decrease in modulus when the PPO is in blocks. Figure 7.4 shows dynamic mechanical data for the same block copolymer. The glass transitions of both the PPO and nylon 6 are clearly visible.

Figure 7.5 shows the influence of PPO concentration on modulus and impact strength. As with urethanes, a wide range of properties can be obtained by varying PPO level; 20% is most common in nylon RIM because of its combined high impact strength and modulus.

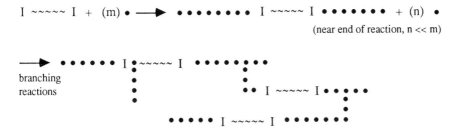

Figure 7.3   Schematic representation of the polymerization of caprolactam (shown as •) from an oligomeric di-initiator to form an ABA block copolymer. At high conversion branching reactions can occur from amide groups in the chain or from ester groups next to the initiator.

Table 7.3 Comparison of Mechanical Properties Between Different Types of Rubber Modified Nylon 6[a]

	% PPO	Morphology	Notched Izod kJ/m^2	Flex Modulus MPa
homopolymer	0	homog.	4	3250
blend	20	blobs >5μm	5	2420
blend + block	20	spheres 1μm	12	2310
block copolymer	20	lamellar 3μm	54	1370

(a) van der Loos and van Geenen (1984, 1985)

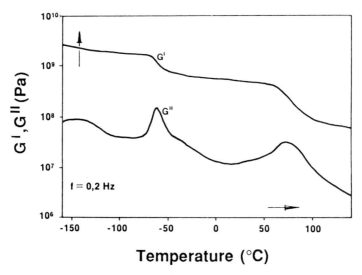

Figure 7.4   Dynamic shear storage G' and loss G" moduli vs temperature for nylon 6 block copolymer, 20% PPO (van der Loos and van Geenen, 1985).

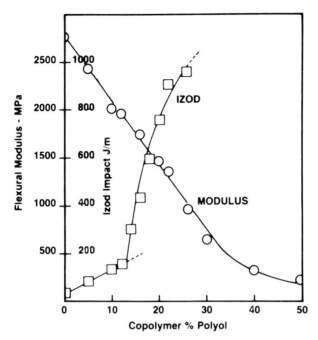

Figure 7.5 Variation of fluxural modulus and impact energy with polyol block content in nylon 6 copolymer (Gabbert and Hedrick, 1986).

Table 7.4 gives typical formulations for nylon RIM based on the two different initiators shown in Figure 7.2. Trifunctional PPO of $M_n \approx 5000$ is often used to promote a lightly crosslinked final product. Other soft segment oligomers like hydroxy terminated polybutadiene can be used. The MgBr lactam formulation is similar to Nyrim PI-2000 developed by Monsanto (Hedrick, Gabbert and Wohl, 1985) and the Na lactam to DSM patents (van Geenen et al., 1985).

The kinetics of anionic polymerization have been well studied. Malkin and coworkers (1982) found that an autocatalytic model fit a large amount of data for the sodium catalyzed reaction with both acyl and isocyanate based initiators. In terms of

$$d\alpha/dt = k \exp(-E_a/RT) \cdot (I^2/M_0) \cdot (1 - \alpha) \cdot (1 + b\alpha/I) \qquad (7.1)$$

conversion α where I is the concentration of initiator and was always the same as the catalyst; M_0 is the initial concentration of monomer. There are three constants in the model:

the activation energy, E_a = 63 ± 6 kJ/mole for all the results, k is the front factor reflecting reaction speed and b is the autocatalytic term. Both depend on the type of initiator chosen. They found 1,6-hexane diisocyanate, HDI, to be the fastest reacting initiator.

Table 7.4   Typical Nylon RIM Formulations

	TANK A 90° C		TANK B 90°C	
MgBr[a]	caprolactam	60%	caprolactam	99%
	PPO-Acyl	40%	MgBr lactam*	~1%
Na[b]	caprolactam	60%	caprolactam	99%
	PPO-Iso	40%	Na lactam*	~1%

(a) Hedrick, Gabbert and Wohl (1985)
(b) van der Loos and van Geenen (1984, 1985)

*Typically the molar concentration of initiator groups is equal to catalyst and both are about 0.2 mol/L in the final mixture. Flow ratio of tank A to B is 1:1.

Table 7.5   Kinetic Constants for Polymerization of Caprolactam with Sodium and HDI

	T_o °C	M_o mol/L	I [1] mol/L (of NCO groups)	E_a kJ/mol	k L/mol s x 10^{-8}	b	ΔH kJ/mol
homopolymer [2]	131 139	8.84	0.18	63.8±.5	2.23±.1	1.15±.5	16
block copolymer [2]	126	7.42	0.15	63.8	2.23	1.15	-
Malkin et al. (1982)	160	~8	<0.1	63±6	4.17	0.066	15

(1) Concentration of sodium caprolactam = isocyanate
(2) Results of adiabatic temperature experiments from Sibal, Camargo and Macosko (1984), Sibal (1982).

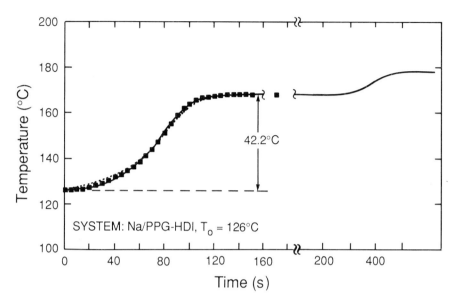

Figure 7.6   Adiabatic temperature rise due to anionic polymerization of caprolactam followed by crystallization of the nylon 6. Sodium caprolactam cayalyst, HDI capped PPO Mn ≅ 2000 as the diinitiator, 16 wt%. Dotted line is the temperature rise predicted from the homopolymerization kinetic constants, Table 7.5.

Sibal, Camargo and Macosko (1984) found Malkin's model fit their data for both homopolymerization and with 16% PPO using sodium caprolactam and HDI initiator. Figure 7.6 shows the adiabatic temperature rise data for their block copolymer formation. The dotted line is calculated from the kinetic constants determined for the homopolymerization. The values of the constants are given in Table 7.5. Their value of activation energy is in good agreement with Malkin, et al., and that for the front factor is reasonable. However, their auto acceleration term is much higher. This may be due to Sibal's lower initial temperatures, which are closer to those used in nylon RIM. The infrared data of Ishida and Scott (1986) indicate that crystallization occurs simultaneously with reaction at 160°C for caprolactam homopolymerization. This crystallization could accelerate the reaction.

At longer times, Figure 7.6 also shows the exotherm due to crystallization. Crystallization is essential to the solidification of the polymer so that it can be demolded. Although there are no detailed studies of nylon 6 crystallization during polymerization, crystallization rates are available on already made polymer. Figure 7.7 shows the data of Magill (1962) which indicates a maximum crystal growth rate at 135°C.

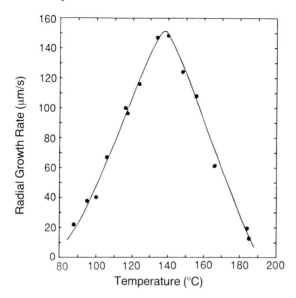

Figure 7.7   Crystal growth rate vs temperature for nylon-6 homopolymer (Magill, 1962)

To better understand nylon RIM it is helpful to compare its process requirements to those given in Table 7.2. Storage stability is a greater concern than for polyurethanes since the catalyst slowly polymerizes caprolactam even without the initiator. Catalyzed caprolactam can be stored for up to 24 hours. Water vapor discolors the reactants and CO_2 can deactivate the catalyst so the tanks should be blanketed with dry nitrogen. Stainless steel should be used to prevent iron contamination. Reactants can be fed as solids to the conditioning tanks. However, for large tanks melting is too slow so premelted caprolactam is generally used.

Conditioning must be done above the monomer melting temperature, (67°C); typically 90°C is used. This requires heated tanks and all flow lines to be heat traced, which means a specially designed high temperature RIM machine (Hall, 1984). The gas dispersion problem seems simpler than for polyurethanes since dry nitrogen dissolves readily in caprolactam. Quantities sufficient for mold packing, at least 20 vol.%, will dissolve in the conditioning tank under 3 bar pressure at 90°C (Kavanaugh, 1987).

Metering and mixing steps also seem to be easier than for polyurethanes. The polymerization is chain-wise with monomer in both reactant streams, so tight stoichiometric control is unnecessary. Since reactant viscosity is low and both streams are completely compatible, low pressures, ≤ 5 MPa, are sufficient to produce good mixing.

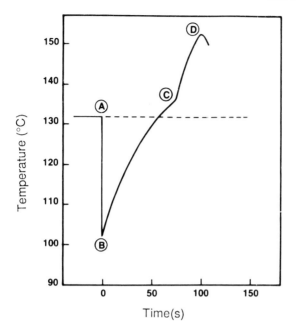

Figure 7.8   Temperature vs time near the center of a 3 mm thick plaque mold during the curing step with Nyrim PI-2000, T_{mold} = 132°C. The rise from B to C is due to the reaction exotherm and from C to D due to crystallization (Gabert and Hedrick, 1986).

The low viscosity and relatively slow reaction mean that molds can be easily filled. In fact since gel times are typically over 20 s, two shots can be used to fill large molds. However, the low viscosity may lead to bubble problems. Similar to the results with polyurethanes (see Figures 5.5, 5.29 and 6.11), Kavanaugh (1987) reports that large bubbles become a problem for Nyrim PI-2000 with high velocity into the mold and high amounts of dissolved nitrogen. He found that when the product of velocity and storage tank pressure exceeded 5 (m/s)(MPa) large voids became a serious problem. Gabbert and Hedrick (1986) report that if they raise the viscosity of the reactants above 0.5 Pa·s by the addition of 1% of a viscosifying agent they can eliminate bubble problems. Sibal, Camargo and Macosko (1984) give isothermal viscosity rise data for the isocyanate initiated homopolymerization.

Figure 7.8 shows the temperature changes during the curing of a nylon RIM plaque. The thermocouple initially reads the mold temperature of 140°C. Temperature drops sharply when the cooler liquid reaches the thermocouple during the filling step, but rises quickly due to reaction. The second rise is due to crystallization. Its onset is proba-

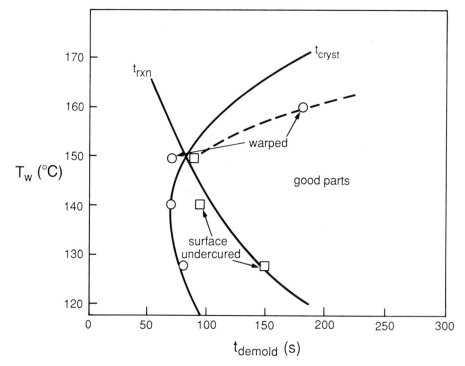

Figure 7.9   Demold time vs mold temperature for a 3mm thick plaque Nyrim PI-2000. At times less than the circled points the plaques were warped. At times less than the square points vapor was visible on demolding indicating ≥ 4% free monomer. Solid lines calculated, see text (Kavanagh, 1987).

bly a convenient gel point and the peak may relate to demold time. However, we need to be cautious with this interpretation since the mold is neither adiabatic nor isothermal, and the thermocouple location is not known precisely and especially since we do not know yet how crystallization relates to the modulus and tear strength needed for demolding.

However, Kavanaugh (1987) has made some useful demolding observations. At short demolding times he observed two types of defects, warpage and monomer vapor. The latter is due to insufficient reaction. These data correlate very well with time to 96% reaction. The activation energy of Table 7.5 was used with k adjusted to give the best fit since the specific kinetics for this system were not available. Warpage is probably due to incomplete crystallinity on demolding but may also be influenced by the fact that the temperature is above the T_g for nylon 6, ≈ 75°C. The t_{cryst} line in the figure is the time calculated for a spherulite to grow to 1cm at the temperature of the mold surface, T_w, using the growth rate data of Magill. The agreement is good except for $T_w = 160°C$. Time to the peak of the crystallization exotherm like that shown in Figure 7.8 also gives fair agreement at 128-150°C but grossly underpredicts the time at 160°C. More crystallization kinetics information on the block copolymer is needed.

It is useful to compare Figure 7.9 to the general moldability diagram for phase separating polyurethanes, Figure 6.25. They are qualitatively quite similar. At low temperatures reaction rate is slower than phase separation or crystallization, but at high temperatures these physical phenomena slow down. In polyurethanes this is due to increased compatability between the copolymer blocks. In nylon 6 it is due to slower crystallization because of reduced supercooling below its melting temperature. One difference between polyurethane and nylon is that at low temperatures the early crystallization of the nylon does not seem to stop the polymerization as does phase separation for polyurethanes. Sibal (1983) did, however, report that lower mold temperatures yielded more brittle polymers.

Mold release is quite easy with nylon 6. No mold release agents need to be added to the formulation; however, for complex parts mold release is improved if one mold surface is held 5° C lower than the other. Postcuring is not necessary, but as discussed above polymerization must be ≥ 96% complete on demold to prevent monomer release. Parts can be painted without special treatment.

In summary nylon 6 block copolymers can be readily made via RIM. The reactants are easier to mix and much less sensitive to ratio than polyurethanes. The two major processing disadvantages are higher temperature and longer demold times. High temperature RIM machines can solve the first problem but the second is more basic. To make nylon RIM a high speed process, both polymerization and crystallization rate must be increased. Gabbert and Hedrick (1986) report that by increasing reactant temperature to 130°C and using a propietary activator they can reduce demold time to 40s.

Nylon 6 block copolymers have the best combination of modulus, impact strength and use temperature in an unfilled RIM polymer. Because nylon 6 is more hydroscopic than most polymers, they suffer larger dimensional changes and mechanical property variation with humidity changes.

7.2 Dicyclopentadiene

In 1983 Hercules introduced the first polymer which was invented particularly for RIM (Geer, 1983; Klosiewicz, 1983). Polydicyclopentadiene, PDCP, is a crosslinked pol-

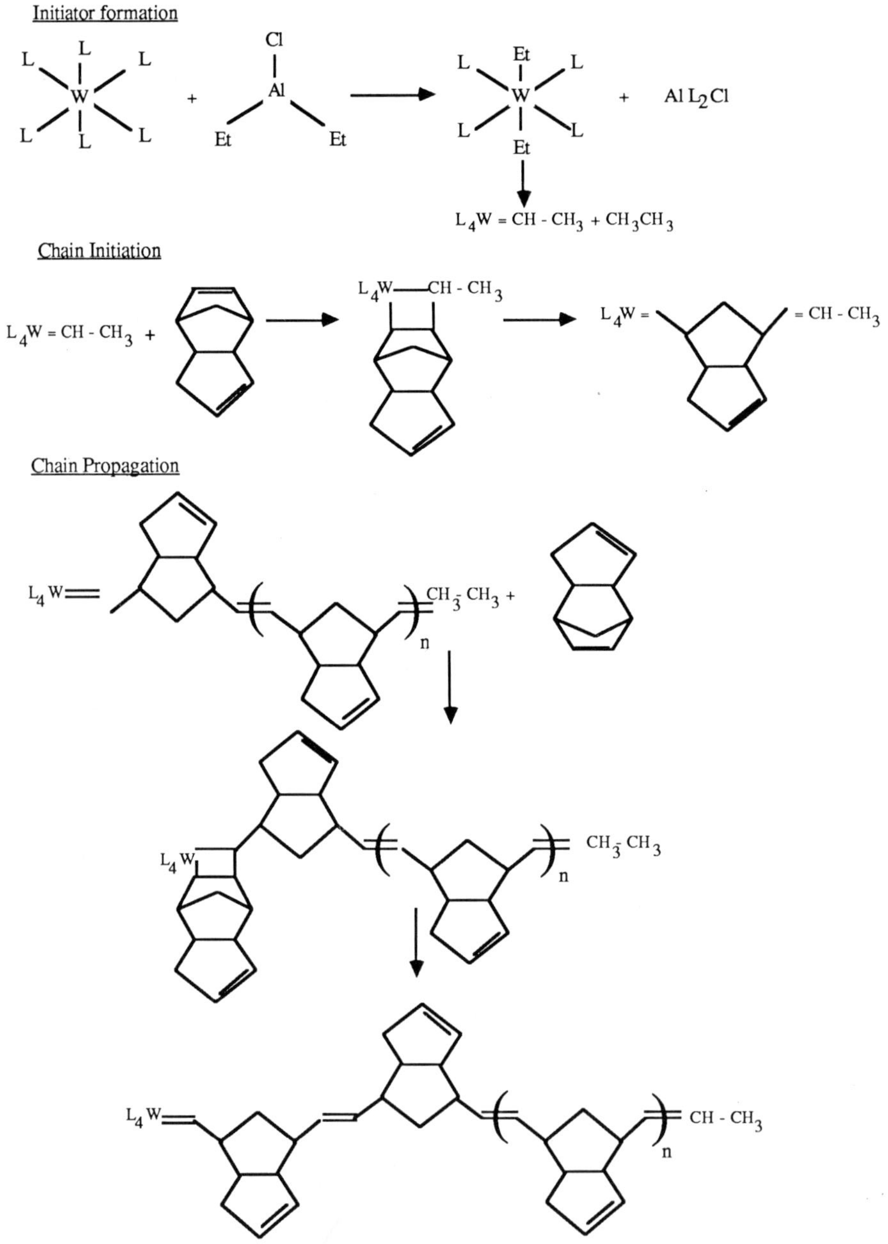

Initiator formation

Chain Initiation

Chain Propagation

Figure 7.10   Steps in dicyclopentadiene polymerization by metathesis; Et = - CH_2CH_3 and L = ligand such as Cl.

ymer which forms very rapidly at room temperature by metathesis of its low viscosity monomer. It is well suited for RIM; in fact its fast crosslinking and the need to exclude oxygen during mixing make RIM the most convenient method to produce PDCP. Besides easy RIM processing it has high modulus along with high impact strength, as illustrated in Table 7.1.

The polymerization steps are outlined in Figure 7.10. The initiator is a coordination type catalyst formed by the reaction of a tungsten or molybdenum compound with an alkylaluminum chloride. These initiators are destroyed by O_2 or H_2O, essentially in direct proportion to their molar concentration.

The initiator opens the highly strained norbornene ring. Propagation is by a metathesis mechanism through the norbornene; less than 5% of the double bonds are consumed (Matejka, Houtman and Macosko, 1985). The less strained cyclopentene ring can also react leading to the crosslinked structure illustrated in Figure 7.11. In typical PDCP RIM it appears that about 20% of these rings are opened. A two part RIM system can be formulated by putting the WCl_6 in one tank with DCP monomer and the Et_2AlCl plus DCP in the

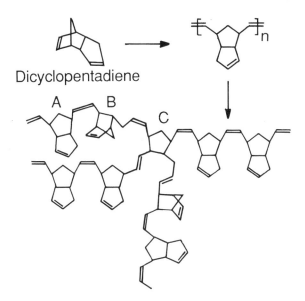

Dicyclopentadiene

Figure 7.11 PDCP structure: (unit A) polymerization through the norbornene ring; (unit B) polymerization through the cyclopentene ring; (unit C) crosslink or branch due to polymerization through both rings.

other. However, the polymerization is too fast even at room temperature. Klosiewicz (1983) found that ethers, esters or ketones could be used to delay the formation of the initiator; in particular di-n-butyl ether was found to be very effective. In addition WCl_6 must be made soluble in DCP. This can be done by first suspending it in toluene and then reacting it with a phenolic compound like p-tert-butylphenol. This compound will polymerize DCP in a few hours, so a chelating agent such as acetylacetone or benzonitrile is added. This prevents significant polymerization for at least 4 weeks (Klosiewicz, 1983).

All these components are combined as shown in Table 7.6 to create a DCP RIM formulation. As indicated in the table, 5% of an elastomer like Kraton 1102, a styrene butadiene styrene triblock copolymer, is also added. This improves impact strength and increases reactant viscosity for less air entrainment during filling and less flash.

Adiabatic temperature rise data are given in Figure 7.12 for a formulation similar to that in Table 7.6, without the elastomer and with acetylacetone instead of benzonitrile as the catalyst stabilizer. These temperature rise curves have four characteristic features: an inhibition period, a very rapid rise, a second rise and sometimes an endotherm at 180°C.

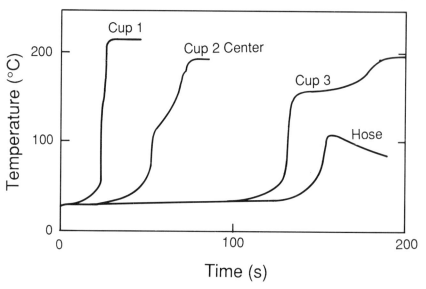

Figure 7.12   Adiabatic temperature rise for a DCP RIM formulation with a) 1% b) 3% c) 10 mole % inhibitor in insulator cups and d) 10% inhibitor in an uninsulated silicone tube about 6 mm diameter (replotted from Matejka, Houtman and Macosko, 1985).

Table 7.6   **DCP RIM Formulation**

TANK A		TANK B	
35°C		35°C	
DCP	94 wt%	DCP	94 wt%
Kraton 1102	5 wt%	Kraton 1102	5 wt%
Et_2AlCl	0.67 mol %	toluene	1.5 wt%
Bu_2O	1.0 mol %	WCl_6/phenol	0.1 mole %
antioxidants		benzonitrile	0.11 mole %

A and B are mixed at a 1:1 ratio; Et = CH_3CH_2- and Bu = $CH_3CH_2CH_2CH_2$- (Klosiewiez, 1983).

The adiabatic temperature rise results and the mechanism of Figure 7.10 suggest a simple, two-step kinetic model for the polymerization of DCP. The first step is an inhibition period in which the di-n-butyl ether or similar compound rapidly complexes with the $L_4W=CH\text{-}CH_3$ as it forms. The decay of this complex Z^* controls the chain initiation

$$Z^* \xrightarrow{k_d} Z_d + M^* \tag{7.2}$$

where Z_d represents some deactivated form of the inhibitor and M^* is the active chain initiator. This zero order decomposition can be written simply as

$$-\frac{dZ^*}{dt} = k_d \tag{7.3}$$

which can be integrated to give the inhibition time

$$t_i = t_o + k_d Z_o \tag{7.4}$$

Here we have assumed $Z^* = Z_o$ the initial concentration of inhibitor; t_o is some initial time for the formation of $L_4W=CH\text{-}CH_3$.

The data of Figure 7.12 indicate inhibition times of 20, 50 and 130s for $Z_o = 1, 3$ and 10 mole %. Thus $t_o = 13s$ and $k_d = 150$ l • s/mol (or 12s/mole %). Klosiewicz (1983) with benzonitrile as the stabilizer found $t_o = 25s$ and $k_d = 260$ l • s/mol. In both studies the polymerization started at 25°C. Klosiewicz indicated that the rate doubles every 8°C leading to $E_d = 68.6$ kJ/mol.

The second step of the simple kinetic model is propagation. If we assume that the initiator concentration is constant and that the rate doesn't change as monomers, M, add to it, we can write

$$M_n{}^* + M \xrightarrow{k_p} M_{n+1}{}^* \tag{7.5}$$

The rate expression becomes

$$-\frac{dM}{dt} = k_p M_n{}^* M \tag{7.6}$$

which can be integrated to give

$$\ln M_o/M = k_p [WL_6] (t - t_i) \tag{7.7}$$

Here we have assumed that all of the tungsten compound becomes active chain initiator, $[WL_6] = M^*$, after the inhibition period. This equation can be fit to the rising portion of the adiabatic temperature curve to give values for $k_p = A_p \exp E_p/RT$.

The work of Matejka, Houtman and Macosko (1985) indicates that the second portion of the adiabatic rise may lead to degradation of properties. The endotherm at 180° may correspond to boiling of residual monomer. Since RIM molds typically have good heat transfer and are less than 6 mm thick, this second rise is not expected to occur (see curve d, Figure 7.12). The first exotherm is about 125°C which leads to a heat of reaction of about 45kJ/mol DCP. This is close to the energy required to open the norbornene ring.

RIM processing of DCP is in many ways simpler than for urethanes. The main complication is preventing the monomer tanks from exposure to water and oxygen. Normal levels of air and moisture in the mold cavity before filling seem to present no problems. Although DCP monomer is not particularly toxic it has a strong olefinic odor. Well sealed RIM equipment and well vented mold areas are used to improve the work environment.

Monomer temperature is confined to a fairly narrow temperature range between 28°C, the monomer freezing point, and 40°C, the flash point. Typically 35°C is used. Dissolving gas to provide mold packing appears to be unnecessary. The early gelation of this chainwise polymerization followed by thermal expansion due to the rapid exotherm seem to adequately compensate for polymerization shrinkage.

Like nylon 6 a 1:1 ratio is used and a wide stoichiometric imbalance can be tolerated. And, also like nylon, low component viscosity, complete component compatibility

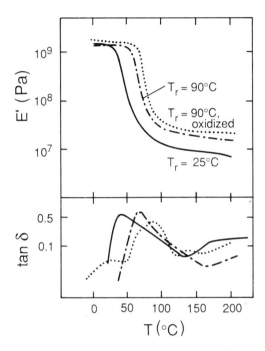

Figure 7.13   Dynamic tensile storage modulus E' and tan δ vs temperature for 0.2 mm films of DCP polymerized at 25°C and 90°C without and with exposure to air for several hours (replotted from
Matejka, Houtman and Macosko, 1985).

and the chainwise polymerization mechanism make mixing easier. The critical Reynold's
number should be lower than for urethanes. Typical impingement mixing pressures are
≤7MPa (1000 psi).

As indicated above, addition of a liquid elastomer is used to control air entrainment
during mold filling. Inhibitor level and mold temperature can control premature gelation.
However, the gel point occurs at very low conversion, $\alpha < 1\%$ according to Matejka, Houtman and Macosko (1985). Geer (1983) indicates that $t_{gel} \cong t_i/2$. This gel will be weak. It
may be possible for such a gel to break and flow and then reheal in the mold.

Since initial temperatures are 35°C while molds walls are held at 50 - 80°, the
polymer will start curing at the wall and then rapidly exotherm. This exotherm drives the
temperature, even at the surface, over the final polymer glass transition, about 110°C, and
thus results in complete reaction in the mold. Residual monomer levels are below 1%.

As the polymer at the surface cools below T_g it releases from metal surfaces and stiffens enough to be readily demolded. No mold release is required, but, as with nylon RIM, a difference in temperature between mold surfaces aids release. Typical mold shrinkage is 1%. The surface polymer oxidizes after several hours exposure to air. This skin of about 50 μm provides for good paint adhesion and helps to seal any residual monomer into the part (Geer and Stoutland, 1985). Because of its olefinic nature DCP has excellent solvent resistance.

Figure 7.13 shows the effect of polymerization temperature and oxidation on dynamic mechanical properties. Properties of the commercial polymer are indicated in Table 7.1. It has high modulus and toughness; the main limitation seems to be low heat distortion temperature. However, because the polymer is highly crosslinked it will deform but not flow at high temperatures. The addition of about 10% oligomerized DCP raises T_g from 110 to about 135°C (Matlack, 1987).

7.3 Other Non-Urethanes

A large number of polymerization chemistries have been tried for RIM. Not much has been described in the literature and none appear to be yet as commercially successful as nylon-6 and DCP. All build structure by crosslinking and develop a high modulus. They are rather brittle, requiring continuous fiber reinforcement to build high impact strength (compare Table 7.1 to 8.5). Each system is described briefly below.

Unsaturated esters - After polyurethane foams and phenolics, the largest volume reaction processed polymer is unsaturated ester-styrene copolymer. One million tons are used per year in the US in sheet molding, bulk molding, injection, transfer, resin transfer, hand lay up, filament winding and pultrusion. All these processes are based on heat activated, free radical initiated polymerization from premixed reactants. Free radical inhibitors are used to achieve storage stability and increase gel time long enough to fill the mold.

It is possible to use the same chemistry to formulate a mixing activated system. With a low inhibitor level a peroxide initiator and a promotor or accelerating agent such as cobalt octoate, gel times as short as 4s have been reported starting with near room temperature reactants (Sardla et al., 1983).

Azo compounds which decompose to N_2 have been added to compensate for the large shrinkage which typically accompanies vinyl polymerizations (Kubiak, 1980). Chopped glass or glass mats, particulate fillers like $CaCO_3$ or $Al_2O_3 \cdot 3H_2O$, and mold release agents like zinc stearate are added.

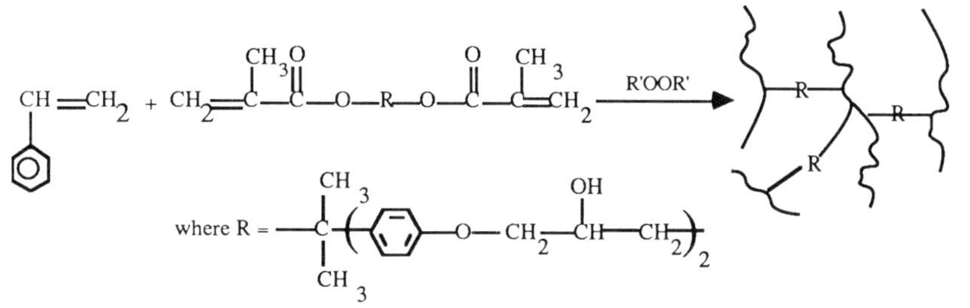

Figure 7.14 Schematic diagram of the free radical copolymerization of styrene with methacrylic ester of bisphenol A-epichlorohydrin.

Figure 7.14 illustrates the copolymerization of vinyl ester with styrene to form a highly crosslinked network. The free radical initiator starts a chain reaction through the double bonds. The divinyl molecules form the crosslinks between chains. Unsaturated esters typically contain 4 or 5 vinyl groups. RIM formulations incorporate 20 to 50 wt% styrene with the rest unsaturated esters. Like DCP the same manner mixture can be put in each tank with the initiator in one tank and accelerator in the other. Since it is a chain mechanism maintaining stoichiometric ratio is not critical. Preventing reaction in the tank containing initiator is a problem. Stability at ambient temperature for a week is desirable (Kubiak, 1980). The resulting polymer is highly crosslinked and brittle but good properties are achieved using a glass mat (eg. Gonzalez and Macosko, 1983).

The polymerization is by a chain mechanism delayed by an inhibitor so the kinetics are similar to DCP. The important steps are inhibition, initiation and propagation. Chain termination may be neglected due to the high degree of crosslinking and diffusion control early in the reaction. Based on kinetic data, mainly from DSC, a number of kinetic models for the polymerization of styrene with unsaturated esters exist (Stevenson, 1980; Lee, 1981; Gonzalez, 1983; Batch and Macosko, 1987). These models could be applied to RIM filling and curing.

Like DCP, since the styrene-ester is a chain polymerization, it gels at very low conversions. Gonzalez-Romero and Macosko (1985) found gel points at less than 1% conversion. They also report a viscosity-conversion relation for their styrene vinyl ester system.

Acrylamate - Ashland Chemical has developed a variation of unsaturated ester RIM (Borgnaes et al., 1983). Reaction occurs in two steps. First two monofunctional unsaturated alcohols (see Figure 7.15) couple with MDI to give an unsaturated ester-urethane. The heat from this reaction plus the peroxide initiator in the isocyanate component (see Table 7.7) starts the second free radical reaction which builds the highly crosslinked polya-

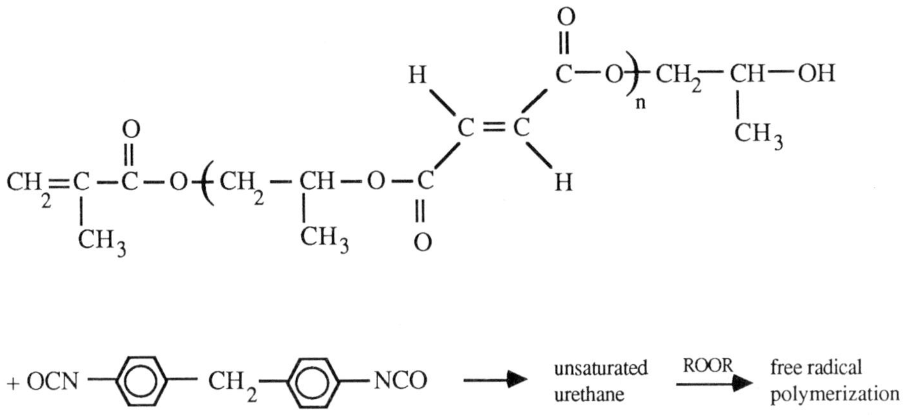

Figure 7.15   Steps in acrylamate RIM system (Borgnaes et al., 1984). The unsaturated monoalcohol is made from methacrylic acid, propylene oxide and maleic anhydride. Two molecules of this combine with MDI to form an unsaturated urethane. The heat from this reaction causes a peroxide to initiate free radical polymerization of the double bonds which leads to a highly crosslinked polymer. n = 0-4 but typically n = 1.

crylamate. Scott (1986) has identified both reactions by infrared. Gel time for the system is 4 to 20s and demold 1 to 2 min at $T_w = 100°C$.

The final polymer has a high modulus and 130°C heat distortion temperature but like the styrene-esters has very low impact strength. When combined with glass mats in the structural RIM process excellent properties and impact strength are obtained.

Table 7.7   Acrylamate RIM Formulation[a]

TANK A 25°C		TANK B 25°C	
unsaturated alcohol	99 wt%	u-MDI	95.2 wt%
DBTDL	0.1-0.25	t-butyl perbenzoate	4.8%
cobalt accelerator	0.1-0.3		

(a) Borgnaes et al., 1984. A and B are mixed at a 2:1 wt ratio. Properties can be obtained as indicated in Table 8.5 (Kelly, 1986).

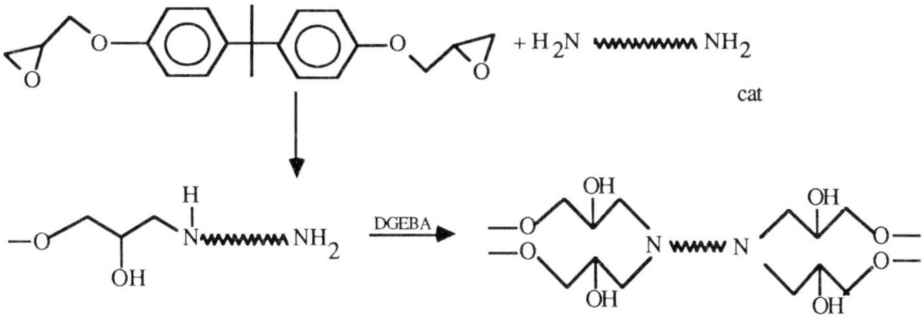

Figure 7.16   Polymerization of diglycidal ether of bisphenol A (2,2-bis(4-hydrophenyl) propane) with 4,4'-di-cyclohexylamino)methane. The epoxy groups react with both primary and secondary amine groups to form a crosslinked network.

Epoxy - By using catalysts and hot molds cast epoxy formulations can be adapted to RIM. Most formulations are based on bisphenol A epoxy resins and aliphatic diamines. Figure 7.16 illustrates the chemistry. Since both amino hydrogens can react with epoxy groups, the diamine molecules are four functional. This leads to a highly crosslinked network.

There are a large number of aliphatic amines which are liquid at room temperature that can be used with liquid epoxies. De la Mare and coworkers (1980; 1983) report RIM studies with the 4,4'-di(cyclohexylamino) methane formulation given in Table 7.8. They found alkali metal salts like calcium or lithium nitrate catalyzed the reaction. The polyethylene oxide seems to be added to stabilize the catalyst. With a 131°C mold and a 90s demold time they obtained the properties given in Table 7.1. They also report results with a polyamide amine (Versamid). Waddill (1980) used methyliminobispropylamine and could remove glass mat reinforced parts in 90s from a 120°C mold. RIM molding can be done under similar conditions but using aminoethylpiperazine or diethylenetriamine (Osinski,1983) or triethylenetetramine (Kim and Kim, 1987). In these formulations no catalyst was added. However, the amino molecules contain secondary and tertiary amines which can be catalytic. Kubiak (1980) used methyltetrahydrophthallic anhydride with bisphenol A and cycloaliphatic based epoxies. These formulations required higher mold temperatures (150°C) and three minute demold times.

Epoxy RIM formulations are typically stoichiometric: one epoxy group per amino hydrogen. However, because the resulting polymer is highly crosslinked and because epoxy can react with the OH groups produced in the amine reaction, stoichiometric balance

Table 7.8   Epoxy RIM Formulation

TANK A		TANK B	
45 - 75°C		30-50°C	
diglycidal ether of bisphenol A (Epon resin, M_n = 350)	100 parts wt.	4,4-di(cyclohexylamino) methane (PACM-20)	28 parts
$Ca(NO_3)_2$	1		
polyethylene oxide (M_n = 400)	1		

A and B are used at 3:3 to 1 ratio (de la Mare and Brounscombe, 1983; de la Mare et al., 1980).

is not as critical as with urethane RIM formulations. An imbalance of ±10% can be tolerated. Epoxy formulations are also not particularly sensitive to oxygen or trace amounts of water.

Standard RIM equipment and pressures have been used for epoxies (de la Mare et al, 1980). Because of the high viscosity of typical epoxy resins, tank temperatures of about 60°C are required. However, because the reaction is slow at mixing temperatures and only activated by the mold heat, impingement mixing is not really necessary. Manzione and Osinski (1983) used hand mixing and Kim and Kim (1987) used a stirring mixhead.

Adiabatic temperature rise and differential scanning calorimetry have been used to measure epoxy kinetics. Osinski (1983) reports simple nth order kinetics fit his thermal measurements on a stoichiometric aminoethylpiperazine formulation above 80°C.

$$dC/dt = 4.977 \exp(-7991/T)\ C^{2.8} \qquad\qquad (7.8)$$

where C is epoxy/m^3. Kim and Kim report a lower order, 1.64, for triethylene tetramine at 70% of the stoichiometric amount. They also find that Equation 2.43 fits their viscosity rise during reaction. Osinski also applied the same equation but needed a conversion dependent flow activation energy to fit his data. Because epoxy reactions are relatively slow compared to typically injection times viscosity buildup and premature gelation are not important problems in epoxy RIM recall Figure 6.26). However, initial viscosity as a function of temperature is important for fiber wet out in the structural RIM applications discussed in the next chapter.

Reaction kinetic expressions like the one given above are valuable for predicting maximum mold temperatures and demold times. Some results from Manzione and Osinski are shown in the previous chapter, Figure 6.26. Because of their large exotherm and the necessity to use hot molds thermal degradation is a real problem in epoxy RIM.

Epoxies offer high modulus, solvent resistance and high usage temperatures. They present no special problems for RIM molders except high mold temperature. But their cost and particularly brittleness seem to have limited their application.

Isocyanurate - In Chapter 2 Table 2.2 we saw that isocyanates can trimerize to form the isocyanurate ring. This reaction is promoted by tertiary amines and bases and can be used to build up a highly crosslinked polyisocyanurate via RIM. Typically about 30% polypropylene polyol of about 1000 equivalent weight is also used to reduce brittleness and the maximum exotherm. Thus the resulting polymer is built from a mixture of urethane bonds and isocyanurate (Carleton et al., 1978). No additional urethane catalyst is needed. The reactions are quite rapid, giving gel times of 10 to 20s and demold in 60s or less (Carleton et al., 1985). Vespoli and Alberino (1983) have modelled the two competing reactions and obtained kinetic parameters using the adiabatic method described in Chapter 2.5.

Even with the added polyol the polymer is still too brittle to be used neat. The main application is as a matrix resin for S-RIM (Carleton et al., 1985). Properties with 38% glass are given in Table 8.5.

The cyanate group can also react to form a similar trifunctional ring to build a crosslinked polycyanurate. Although the reaction is slower than that of isocyanates the resulting polymer is tougher (Shimp, 1987).

Phenolics - Phenolics are the oldest synthetic polymer, yet formulations have been developed for RIM (Forsdyck, 1981; Brode and Hale, 1983). Resole resins can be cured in closed molds with strong acid catalysts. Figure 7.17 gives the proposed reaction. Note that water is a byproduct. It appears that sufficient mold pressure can prevent this water from causing excessive foaming. Brode and Hale propose catalysts like sulfuric acid and maleic anhydride which lead to heat activated formulations. These require 4 to10 minutes to cure at temperatures in the range of 150°C. Forsdyck indicates that cocatalyst systems have been developed which are mixing activated and can cure in less than one minute at 50°.

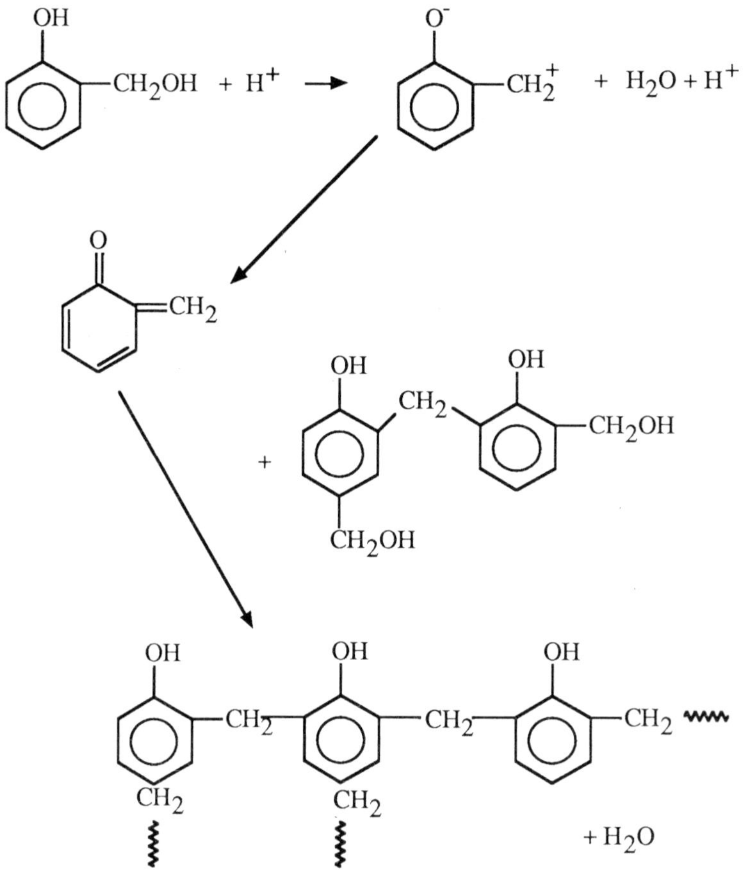

Figure 7.17 Proposed mechanism for acid catalysis of phenolic resin for use in RIM (Forsdyke, 1981).

Phenolics are brittle, therefore, like epoxies and isocyanurates, appear most viable for RIM when reinforced by fiber mats. Phenolics are low cost and have excellant solvent and flame resistance. Resoles can be formulated to low viscosity for good impingement mixing and even for fiber wet out in mat molding, however the acid catalysts mean stainless steel RIM machines and molds. At present there seems to be no commerical phenolic RIM product.

REFERENCES

Alfonso, G.C.; Dondero, G.; Russo, S.; Turturro, A.; "Synthesis, morphology and properties of rubber modified poly(ε-caprolactam) by fast in situ-polymerization," in *Morphology of Polymers*, B. Sedlacek, ed., De Crayter, Berlin, 1986, 427-434.

Batch, G.L., Macosko C.M., "Kinetics of crosslinking free radical polymerization with diffusion-limited propagation", Soc. Plast. Eng., Tech. Papers, *33, 1987*, 974-976

Borgnaes D.; Chappell, S.F.; Wilkinson, T.C., "RIMing of low viscosity crosslinkable alcohols with diisocyanates," Soc. of Plast. Ind., Urethane Technical/Market Conference, 1984 381-286.

Brode, G.L.; Hale, W.F. "Composites made from liquid phenol formaldehyde resins," U.S.Patent 4,403,066, Union Carbide (1983).

Carleton, P.S.; Ewen, J.H.; Reymore, H.E. "High-modulus polisocyanurate elastomers," US Patents 4,126,741 and 4,126,742, 1978.

Carleton, P.S.; Waszeciak, D.P.; Alberino, L.M. "MM/RIM: RIM composites using preplaced reinforcement," Soc. Plast. Ind. Urethane Technical/Market. Conf. 1985, 154-160.

Carleton, P.S.; Waszeciak, D.P.; Alberino, L.M. "A RIM process for reinforced plastics" Reinforced Plastics/Composites Inst., Soc. Plast. Ind. 1986.

De La Mare, H.E.; Brownscombe, T.F.; Gottenberg, W.G.; Overcashier, R.H. "High modulus epoxy RIM systems," Society of Automotive Engineers Conference, Houston, TX, Feb. 1980.

De La Mare, H.E.; Brownscombe, T.F. "Curable epoxy compositions suitable for use in RIM processes," U.S. Patent 4,397,998, Shell Oil Company, 1983.

Dupre,C.R., "???"ACS Polymer Preprints *25 no. 2, 1984*, 296-298.

Farris, R.D.; Overcashier, R.H.; Gottenberg, W.G. "Structural parts from epoxy RIM using preplaced reinforcement" Reinforced Plastics/Composites Inst., Soc. of Plast. Ind. 1982.

Forsdyke, K.L. "Phenolic resins for RRIM" paper #6 presented at Plastics and Rubber Institute of London Conference on Reinforced RIM, February 1981, Solihull, U.K.

Gabbert, J.D.; Garner A.Y.; Hedrick, R.M., "Reinforced nylon 6 copolymers," *Polym. Comp.* 1983, *3*, 196-199.

Gabbert, J.D.; Hedrick, R.M., "Advances in systems utilizing Nyrim nylon block copolymer for reaction injection molding," *Polym. Proc. Eng.* 1986, 4, 359-373.

Geer, R.P., "Poly(dicyclopentadiene): a new RIM thermoset," Proced., Soc. Plast. Eng. National Techn. Conf., Detroit, MI, Oct. 1983.

Geer, R.P.; Stoutland, R.D. "Olefinic reaction molding of large structural parts," *Soc. Plast. Eng. Tech. Papers* 1985, *30*, 1232-1235.

Gonalez, V.M. "Studies of reactive polymer processing with fiberglass reinforcement," PhD Thesis, University of Minnesota, 1983.

Gonzalez, V.M.; Macosko, C.W. "Properties of mat reinforced reaction injection molded materials," *Polym. Comp.* 1983, *4*, 190-195.

Gonzalez-Romero, V.M.; Macosko, C.W. "Viscosity rise during the free radical crosslinking polymerization with inhibition," *J. Rheol.* 1985, *29*, 259-272.

Hall, C.M. "RIM flexes its muscles in seeking out new markets," *Plast. Eng.* 1985, *41* June, 39.

Hedrick, R.M.; Gabbert, J.D.; Wohl, M.H., "Nylon 6 RIM," in *Reaction Injection Molding* J.E. Kresta, ed., Am. Chem. Soc. Symp. Series 270, Washington D.C., 1985, Chapter 10.

Ishida, H.; Scott, C. "Fast polymerization and crystallization kinetic studies of nylon-6 by combined use of computerized micro-RIM machine and FT-IR," *J. Polym. Eng.*, 1986, *6*, 201.

Kavanaugh, D., Monsanto Co., private communication, 1985.

Kelly, W.L., "New RIM composite adds to the designer's arsenal," *Plast. Eng.* 1986, *42*, 29-31.

Kim, D.H.; Kim, S.C. "Engineering analysis of reaction injection molding process of epoxy resin" *Poly. Comp.*, 1987, *8*, 208-217.

Klosiewicz, D.W., U.S. Patent 4,400,340, Hercules (1983).

Kubiak, R.S. "Taking RIM beyond the urethanes," *Plast. Eng.* 1980, *36, March*, 55-61.

Kurz, J.E. "Block size and distribution in nylon-6 RIM," *Polym. Proc. Eng.* **1985**, *3*, 7-24.

Lee, L.J., "Curing of compression molded sheet molding compound," *Poly. Eng. Sci.*, **1981**, *21*, 483-492.

Magill, J.H.; "Crystallization kinetics study of nylon 6," *Polym.* **1962,** 655-664.

Malkin, A.Y.; Ivanova, S.L.; Frolov, V.G; Ivanova, A.N.; Andrianova, Z.S. "Kinetics of anionic polymerization of lactams. (Solution of non-isothermal kinetic problems by the inverse method)," *Polymer* **1982**, *82*, 1791-1800.

Malkin, A.Y.; Beghishev, V.P; Keapin, I.A. "Macrokinetics of polymer crystallization," *Polymer* **1983**, *83*, 81-84.

Manzione, L.T.; Osinski, J.S. "Moldability studies in reactive polymer processing" *Polym. Eng. Sci.* **1983**, *23*, 576-585.

Matejka, L.; Houtman, C.; Macosko, C.W. "Polymerization of dicyclopentadiene: a new reaction injection molding system" *J. Appli. Poly. Sci.* **1985**, *30,* 2787-2803.

Matlack, A.S. "Metathesis polymerization of thermally oligomerized dicyclopentadiene," US Patent 4,703,088, Hercules, (1987).

Osinski, J.S. "Characterization of fast-cure resins for reaction injection molding," *Polym. Eng. Sci.*, **1983**, *23*, 756-762.

Saidla, W.; Peters, G.M.; Mitchell, R.L. "Unsaturated polyester reinforced reaction injection molding and reaction injection molding," Reinf. Plast./Composites Inst., Soc. Plast. Ind. Feb. 1983, paper 23A.

Scott, C.E. "Application of analytical characterization techniques to reaction injection molding and multi-component matrix/rubber/filler polymer composites" MS Thesis, Case Western Reserve University, 1986.

Shimp, D.A. "Thermal performance of cyanate functional thermo-setting resins," *SAMPE Quart.*, **1987**, *19, 41-46.*

Sibal, P.W.; "Anionic polymerization of nylon 6," M.S. Thesis, University of Minnesota, 1982.

Sibal, P.W.; Camargo, R.E.; Macosko, C.W. "Designing nylon-6 polymerization systems for RIM," *Polym. Proc. Eng.* **1983-1984**, *1*, 147-169.

Stevenson, J.F., Processing of reactive materials: A kinetic model for free radical copolymerization," *Soc. Plast. Eng., Tech. Papers.*, **1980**, *26*, 452-456.

van Greenan, A.A.; Bongers, J.J.M.; van der Loos, L.M.; Vrinssen, C.H. "N-substituted carbamoyl-lactam compound," U.S. Patent 4,540,515, DSM (1984).

van der Loos, J.L.M.; van Geenen, A.A. "New developments in RIM nylon block copolymers," presented at 3rd Int. Conf. on Reactive Polymer Processing, Strasbourg, Sept. 1984.

van der Loos, J.L.M.; van Geenen, A.A. "Properties and morphology of impact modified RIM nylon", J.E. Kresta, ed. "Reaction Injection Molding," Am. Chem. Soc. Symp. Series 270, Washington D.C., 1985.

van der Loos, J.L.M.; van Geenen, A.A. "Relation between composition, processing and mechanical properties of RIM nylon block copolymers," presented at Symposium on Reactive Polymers, Naples, June 1986.

Vespoli, N.P.; Alberino, L.M. "Computer modeling of the heat transfer processes and reaction kinetics of urethane-modified isocyanurate RIM systems," *Polym. Proc. Eng.* **1985**, *3*, 127-147.

Waddill, H.G. "Reaction injection molding (RIM) with epoxy resin systems," 35th Annual Conference, Reinforced Plastics/Composites Inst., Soc. Plast. Ind., 1980, paper 22B.

8

REINFORCED RIM

Over half of all RIM materials are filled. The main motivations are to improve dimensional stability and for better mechanical properties. In Chapter 2.4 we mentioned the addition of polymeric particles to polyols. Here we will concentrate on mineral fillers, particularly glass: milled fibers, flakes and continuous fibers. In Chapter 3 we mentioned some problems using filler in RIM equipment. These will be discussed further here.

Polymers typically have a coefficient of linear thermal expansion, CLTE, nearly ten times that of steel. A plastic part 1m long can change nearly 1cm relative to steel in going from summer to winter temperatures. These dimensional changes could cause a polymeric automobile fender to jamb against the door or even tear away from its mounting on the steel frame.

The addition of glass or mineral fibers can greatly reduce the linear expansion of polymers. As shown in Figure 8.1 the CLTE of a typical RIM polyurethane drops more

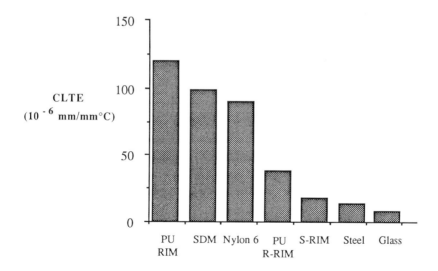

Figure 8.1   Comparison of the CLTE for typical RIM polyurethane, styrene dimethacrylate, nylon 6, R-RIM polyurethane (10 vol% of 1.6 mm milled glass fibers), S-RIM (styrene dimethacrylate system injected into 10 vol% continuous fiberglass mat), steel and glass (adapted from Gonzalez and Macosko, 1983).

than 60% with the addition of 22wt% (10vol%) of milled glass. The Figure shows that continuous glass fibers are even more effective in reducing CLTE. It is important to note that the fibers cannot change the overall volume expansion of the polymer with temperature; they just redirect it. Due to the fiber orientation, thermal expansion in the plane of the part is reduced but increases in the part thickness direction.

Even in applications where thermal expansion is not a major problem, glass fibers help to improve dimensional stability and warpage. Fascia (bumper covers) are attached at the ends of the automotive frame and thus can expand freely away. However, the addition of 10 to 15wt% milled glass fibers prevents waviness in these long thin parts.

The other motivation for adding reinforcement is to improve mechanical properties. The main goal is to increase modulus especially at high temperature (to survive paint oven cycles) without sacrificing impact. We saw in Chapter 2.7 that the modulus of polyurethanes can be increased by raising the hard segment content. This, however, typically leads to lower impact strength.

Reinforcement with long fibers can increase both modulus and impact. But it is difficult to pump such long fibers through RIM equipment. Furthermore, flow in the mold leads to excessive orientation, warpage and poor surface quality. Because of these problems, currently all formulations use short glass fibers or flakes. They sacrifice impact strength for the improved dimensional stability and modulus. Such products might better be called "filled" rather than reinforced RIM (R-RIM).

Another approach is to put continuous fiber mats directly in the mold cavity and inject a low viscosity, reactive mixture through the mats. This process is only just becoming commercial and goes by various names: mat molding RIM (MM-RIM), composite RIM, structural RIM (S-RIM) or fast resin transfer molding (RTM).

This chapter describes these two approaches for reinforcing RIM: the addition of particles or short fibers into the liquid reactants before injection and the preplacement of fiber mats in the mold cavity. Both properties and special processing considerations will be discussed, primarily with respect to RIM polyurethanes and polyureas.

8.1 Filled RIM

Four types of fillers have been studied extensively for use with RIM polyurethane formulations:

Chopped glass - Glass strands are made by bundling together hundreds of tiny glass filaments (10 to 20μm dia) with an adhesive. These strands can be chopped into fairly uniform lengths as indicated in Figure 8.2 and 8.3. When the chopped strands are

Figure 8.2   Scanning electron micrographs of 1.6mm milled glass fibers (left) and 3.0 mm chopped strand glass fibers (right) (courtesy of Taylor, 1988).

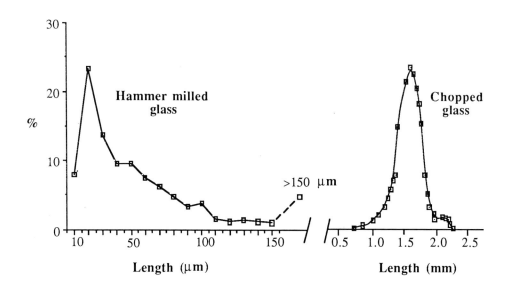

Figure 8.3   Fiber length distribution for hammer milled glass and chopped strand glass. (replotted from Taylor, 1988; Coates et al., 1987).

dispersed in polyol the adhesive dissolves and the filaments separate. Because of their high aspect ratio (l/d~100) these filaments greatly increase the viscosity of the mixture eventually becoming a solid-like mass. Cross et al. (1985) give a relation for the maximum volume fraction of fibers as a function of l/d. Coates and coworkers (1987) found that 16 wt% in the polyol was the maximum concentration at which they could process 1.5mm chopped fibers (ICI WX6450, 17μm dia) in their RIM machine. These suspensions are very shear thinning. At the low shear rates encountered in emptying storage tanks and in recirculation they are very difficult to move but at the high shear rates of impingement mixing and injection they thin down to close to polyol viscosities (Coates et al., 1987). One approach to reduce viscosity would be to chop the fibers shorter but this does not appear to be economical. Another is to use an adhesive that only dissolves at the higher temperatures encountered during and after injection (Girgis, 1985; 1982).

Hammer milled glass - Milled glass is made by hammering glass filaments (~10-20 μm dia) through a screen with a given opening. This opening represents the largest size fiber that will be present in the mixture. The length distribution is very large as shown in Figures 8.2 and 8.3. For what is called 1.6mm (or 1/16in) hammer milled fiber the average length is actually only about 0.2mm. Because the aspect ratio of milled fibers is low they can be loaded to at least 40wt% in polyols and are much easier to disperse and to pump than chopped fibers.

Flake glass - Glass flakes are made by shattering very large diameter, thin walled glass bubbles and passing the pieces through a sieve. The sieve opening size is used to designate the product; 0.4mm (1/64in) is most frequently used in RIM. The flakes have a wide size range as can be seen in Figure 8.4.

Mica - Mica is a plate-like mineral (muscovite, $K_2O \cdot 3Al_2O_3 \cdot 6SiO_2 \cdot 2H_2O$). As with flake glass mica particles are segregated by sieving; 74μm (200 mesh) has been used in RIM studies. These particles are thus on the average smaller than the glass flakes, but they are more regular in shape as illustrated in Figure 8.4.

Other - Other minerals have been tried to reinforce RIM systems. In particular wollastonite ($CaSiO_3$) which is a naturally rodlike particle with about 10:1 aspect ratio has been used (Gerkin et al., 1979). Even talc and calcium carbonate which are more spherical and much softer particles do improve modulus when added to RIM polyurethanes (Gerkin et al, 1979). Titanium dioxide and carbon black are added in small quantities for pigmentation rather than reinforcement.

Table 8.1 compares the important fillers used in RIM. Mica is more dense and smaller than the glass fillers but is much cheaper. Chopped glass is the only one with a

Figure 8.4   Scanning electron micrographs of 0.4mm flake glass flake (left) and mica (right) (courtesy of Taylor, 1988).

Table 8.1   Properties of Fillers for RIM[a]

	Relative cost	Specific gravity	l [b] μm	[b] l/d	Viscosity [c] Pa · s
Polyol	0.75	1.0	--	--	0.2
1.6 mm hammer milled	1.00	2.54	50	5	50
0.4 mm glass flake	1.25	2.54	200	~40	600
200 mesh mica	0.2	2.9	~100	~40	50
1.5 mm chopped glass[d]	NA	2.54	1500	88	2500
Wollastonite	0.5	2.8	~100	10	--

(a) adapted from Ferrarini and Cohen (1982).
(b) l is average length of the fibers (Figure 8.3) and average plate diameter for the glass flake and for mica (Figure 8.4); d is the thinnest dimension.
(c) low shear rate (Brookfield viscometer) at ambient temperature with 38 wt% filler.
(d) 9.3 wt % chopped glass plus 18.7 wt % milled in polyol, strongly shear thinning.

high aspect ratio but its viscosity enhancement is very high even when blended with milled glass.

Ferrarini and Cohen (1982) have compared these four fillers in the same high modulus RIM formulation. Figure 8.5 shows some of their physical property data. All four fillers increase flexural modulus about the same amount, about threefold. This increase in stiffness also translates to much less heat sag. In both properties we note that the fibrous fillers have much more anisotropy; there is a large difference (~1.7 times) between samples cut parallel to the mold filling direction and perpendicular. This is even more apparent in the coefficient of linear thermal expansion. In contrast plate-like particles will orient flat to the flow and thus properties with these fillers will show little dependence in the xy plane. The anisostropy of the fibrous fillers also manifests itself as warpage of large flat panels (0.6m x 1.2m x 3mm).

Figure 8.5d shows that samples with chopped fibers have the greatest tensile strength. This is because they are long enough to break rather than pull out of the matrix. Chisnall and Thorpe (1980) have calculated that for this modulus matrix the fibers need to be at least 500μm long and have good adhesion to the matrix to get true reinforcement.

Both plate-like fillers, glass flakes and mica show lower tensile and impact strength than the fibrous fillers. Rice and Dominguez (1983) also report lower puncture impact energy with platy fillers, especially mica at low temperatures. This is one reason that flake glass has been preferred over mica. We again note in Figure 8.5d that the tensile strength of flake glass is the same in both mold directions and slightly higher than either chopped or milled fiber when measured perpendicular to the filling direction. Coupling agents have little effect on overall properties with flake or milled glass, presumably because the particles are so small (Oertel, 1985, Chapter 7.5). However. silanes such as amino propyl or epoxy propyl silane are used to prevent particle delamination near the surface.

Despite their high reinforcement potential, problems with warpage, anisotropic properties, high viscosity in processing equipment and higher cost have kept chopped fiber from seeing much use. Milled fibers are used in long thin parts such as fascia where warpage is not as critical. Glass flakes are used for large flat parts like fenders where warpage control is critical. Table 8.2 shows properties of a typical polyurea formulation at various flake glass levels. Addition of 20 wt% gives an optimum between modulus and CLTE improvement and impact strength reduction. This filler level is also near the limit of what can be processed conveniently.

In urethane formulations fillers are typically only added to the polyol. Filler surfaces have a tendency to catalyze the isocyanate reactions discussed in Chapter 2.1. Most fill-

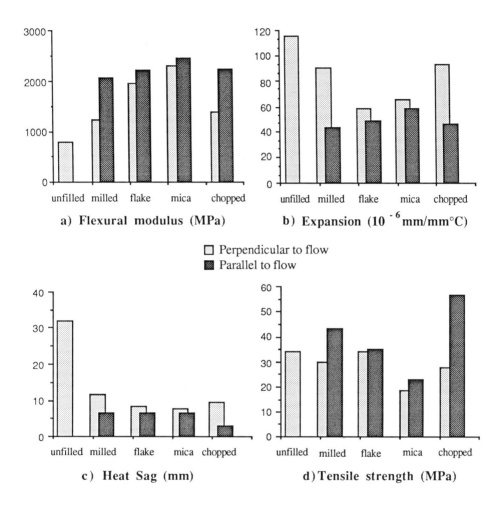

a) **Flexural modulus (MPa)**

b) **Expansion (10^{-6}mm/mm°C)**

☐ Perpendicular to flow
▨ Parallel to flow

c) **Heat Sag (mm)**

d) **Tensile strength (MPa)**

Figure 8.5   Properties of a high modulus polyurethane RIM formulation reinforced with various fillers: chopped glass (5wt%, 1.6mm chopped plus 10 wt%, 1.6mm milled glass); mica (22 wt%); flake glass (22 wt%, 0.4mm); milled glass (22 wt%, 1.6mm) (replotted from Ferrani and Cohen, 1982; see Table 8.1 for more details on formulations).
a) flexural modulus at 23°C;
b) coefficient of linear thermal expansion from -30 to +30°C;
c) heat sag at end of 152mm cantilever after 1 hour at 121°C;
d) tensile strength at 23°C.

Table 8.2   Properties of Polyurea with Flake and Hammer Milled Glass[a]

	Flake[b]					HMG[c]
Wt % Glass	0	10	15	20	25	24
Specific Gravity	1.06	1.13	1.15	1.21	1.23	1.22
E_f (MPa at 25°C)	540	810	1030	1630	1860	1570
E_f (-29°C)	3.0	3.1	3.8	2.9	2.4	3.0
E_f (70°C)						
Tensile Strength (MPa)	24	22	24	33	28	26
Elongation (%)	180	88	47	30	22	10
Izod Impact (J/m)	590	210	200	180	150	280
Heat Sag (mm)						
135°C	16	14	6	5	3.8	3.8
163°C	Fail	56	23	10	9.7	10

(a) Martinez and Vanderhider (1987)
(b) Dow Chemical Spectrim HF with 0.4mm flake glass; 103 isocyanate index; 71°C mold, 121°C post cure for 60 minutes. Mobay Bayflex 150 has similar properties at 20% flake.

ers are hydroscopic and must be dried before introduction into the reactants. The high viscosity of filled reactants makes it necessary to use moving cavity or concentric screw pumps, such as Moyno type, for recirculation. This recirculation as well as tank stirrers prevent the denser fillers from settling out. Lines and fittings must be selected to avoid dead spots and sharp corners. The maximum amount of 1.6 mm milled glass or 0.4 mm flake glass that can be dispersed in polyol and handled in present RIM equipment is about 40 wt%. When mixed with the usual amount of isocyanate this translates to about 25 wt% in the fluid part.

As mentioned in Chapter 3 the only high pressure metering units which can handle these abrasive fillers are displacement pumps, particularly lance pistons. These are normally vertical as indicated in Figure 3.1 and 3.2. Abrasive fillers accelerate wear in mixhead nozzles and can cause jambing of the clean out piston (note particularly Figure 4.6). However, impingement mixing itself appears to obey the same criterion as for unfilled systems, that is a critical Reynolds number, Re $\geq$ 300. In applying the Re criterion the viscosity should be evaluated at high shear rate, $\geq 10^5 s^{-1}$ (Tucker and Suh, 1980). At such high shear rates shear thinning is nearly complete and the viscosity is only slightly higher than that for the polyol. For example, Coates et al. (1987) report polyol viscosity only increased from 1 to 1.35 Pa·s with 16 wt% chopped strand glass fibers at $10^5 s^{-1}$.

At the lower shear rates of mold filling viscosity will be higher and may reduce air entrainment. However, the major problem in mold filling with formulations containing fillers is orientation. Long thin particles line up with the flow leading to the anisotropic properties and warpage discussed above. Plate-like particles also lineup in the shear planes, but, since they are roughly disks or squares, properties are isotropic in the shear plane. Thus, plate-like fillers are preferred for large flat, thin parts.

In the curing step of RIM the maximum exotherm will be lower because filler reduces the concentration of reactive groups. The higher viscosity of filled systems may reduce flash. However, fillers are abrasive and seem to scrub away mold release especially from high shear regions such as near the gate. The higher modulus of filled systems can also mean faster build-up of green strength. The final part may suffer surface finish problems especially with large filler particles. Table 8.3 summarizes the special considerations for filled RIM processing in terms of the unit operations given in Figure 1.2 and Table 7.2.

Table 8.3 Special Considerations in Processing Filled RIM

Supply	• filler must not contaminate reactants (esp. water)
	• high viscosity at low shear
	• moving cavity recirculation pumps
Condition	• settling problems in tanks and lines
	• use tank stirrers
Meter	• only lance pistons
Mix	• use viscosity at high shear to calculate Re_c
	• wear on nozzles
	• jambing clean out piston
Fill	• shear thinning of filled systems may reduce flow instabilities
	• fiber orientation
Cure	• filler reduces T_{max}
Demold	• filler may remove mold release
Finish	• harder to achieve glossy surface

8.2 Fiber Mat Reinforced RIM

In the previous section we saw that the best properties are achieved with long fibers but because of flow orientation the parts produced are very anisotropic. A solution to this problem is to put the fibers in the mold first then inject reactive monomers into them in a second step. This process, generally called structural RIM (S-RIM), is a unique way to produce high performance composites rapidly and economically.

Figure 8.6 illustrates the process. A fiber preform is placed into the empty mold cavity. Pressure is needed to close the mold rapidly and compact the fibers. A normal RIM mixhead is mounted on the mold cavity to inject the reactive monomer mixture. Director radial gates are typically used to prevent displacement of the preform. Polymerization is rapid in the heated mold and parts can be ejected after about two minutes.

The structural RIM process is similar to resin transfer molding (RTM). The key difference is that S-RIM uses a RIM machine to fill the mold and thus high pressure impingement mixing to activate the reaction. RTM typically uses slower reacting, heat activated formulations and inject under low pressures. Components are premixed or flow through static mixers into the mold. Molds are made of lower cost materials like epoxy and held together by manual clamps. Cycle times are 10-60 minutes. S-RIM is aimed for higher volume production. Molds are made of steel and mounted in hydraulic presses. Table 8.4 shows a comparison of RTM to S-RIM for an isocyanurate resin.

The advantage of continuous fiber reinforcement becomes clear if we compare Figure 8.5 and Tables 8.2 to Table 8.5. First the glass contents are higher; S-RIM can easily exceed the maximum for filled systems (~25 wt%) because there is no need to pump the glass into the mold. The fibers are already there. Glass fiber loadings of 60 wt% and higher have been reported (Gonzalez and Macosko, 1983; Slocum et al., 1986). Because of the high glass level and because long fibers can be used the flexural modulus and tensile strength of S-RIM plaques are about five times that of filled RIM. These high values approach those for laminated composites and metals and thus open up load bearing (structural) applications for RIM.

Mechanical properties of fiber reinforced composites can be correlated to fiber content with mean property relations. Gonzalez and Macosko (1983) have done this for the shear modulus of S-RIM samples with a styrene dimethacrylate matrix

$$G \; = \; \varepsilon \, G_g \, (1 - \phi) + G_r \, \phi \qquad\qquad (8.1)$$

where G_g is the shear modulus of the glass fibers, G_r the modulus for the resin matrix, ϕ is the volume fraction of the matrix and ε is an efficiency factor which depends on rein-

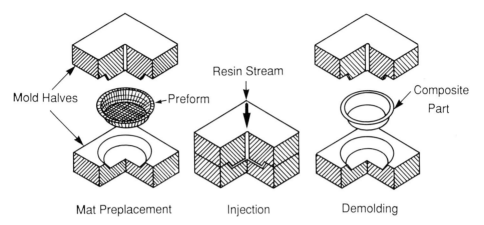

Figure 8.6   Schematic of the structural RIM process. a) Placement of fiber preform in the mold cavity.
b) Closing of mold and rejection of the reactive mixture. Curing in closed mold. c) Opening
of the mold and part ejection (adapted from Gonzalez-Romero and Macosko, 1985).

Table 8.4   Comparison of Structural RIM to Resin Transfer Molding[a]

	RTM	S-RIM
Equipment Cost	$30,000	$500,000
Flow Rate (kg/min)	2.3	55
Mixing	static mixers	impingement
Mold Pressure (MPa)	0.3	2.4
Void Content (vol%)	0.1 - 0.5	0.5 - 2.0
Mold Materials	epoxy	steel
Mold Temperature [b] (°C)	25 - 40	95
Component Viscosities (MPa·s)	100-550	< 200
Cycle Time (min)	10-60	2-6

(a) P.S. Carleton, Dow Chemical, personal communication, 1987.
(b) for Dow's MM 353 isocyanurate system. Other chemistries will differ.

forcement orientation. For random mats ε takes the value 3/8. Figure 8.7 compares this relation to experimental results at two different temperatures. Flexural modulus data for glass mat reinforced polyisocyanurate at ambient temperature also follow Equation 8.1 (Carleton et al., 1985). Slocum et al. (1986) and Nelson (1987) give modulus and strength data as a function of glass loading for the urethane and isocyanurate systems in Table 8.5. They find roughly linear increase in properties up to 50 wt% glass.

Table 8.6 gives typical processing conditions for the four systems described in Table 8.5. The viscosities are somewhat lower and gel times longer than for typical RIM systems. This aids penetration into the fiber mats. The short demold times make S-RIM attractive especially when compared to RTM and other labor intensive composites processes. Mold temperatures and pressures are similar to RIM and thus standard RIM equipment can be used for S-RIM. The additional steps are concerned with getting the fibers into the mold as indicated in the unit operations in Figure 8.8. To compress the fiber mats S-RIM requires mold clamp pressures of about 1MPa, about double that for RIM.

Table 8.5   Typical Properties of S-RIM Composites

	Isocyanurate[a]	Urethane[b]	Acrylamate[c]	Epoxy[d]
Random Glass Mat (wt%) (3 mm thick part)	38	44.8	40	40
Specific Gravity	1.54	1.53	1.46	--
Void (vol%)	1.5	1.5	--	--
E_f (MPa at 25°C)	8,100	9,600	8,700	9,200
Tensile Strength (MPa)	150	150	125	160
Elongation (%)	7.3	2.0	2.1	1.2
Izod Impact (J/m)	510	660	790	~800
Heat Distortion (°C)	184	189	240	>200
Thermal Expansion (m/m°C) x 10-6	~20		27	18

(a) Dow - MM 353 product literature 1986, Nelson, (1987); OCF 8610 mat.
(b) Mobay - Slocum et al. (1986); OCF 8610 mat.
(c) Ashland - Arimax 1100, product literature 1986.
(d) Shell - Farris et al. (1983).

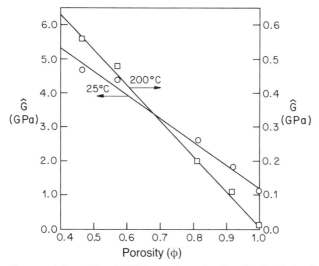

Figure 8.7   Shear modulus at 25 and 200°C vs. volume fraction of resin, ϕ, for chopped strand glass mat (AKM from PPG industries) in a styrene dimethacrylate matrix. Lines from Equation 8.1 using $\varepsilon = 3/8$; $G_g = 22$ GPa at 25°C and 2.8 at 200°C; $G_r = 1$ GPa at 25° and 0.005 at 200°. The highest loading, 45 vol%, resin represents about 70 wt% glass (adapted from Gonzalez and Macosko, 1983).

Table 8.6   Typical Processing Conditions for S-RIM Resins

	Isocyanurate[a]	Urethane[b]	Acrylamate[c]	Epoxy[d]
Viscosity at T_O				
A (mPa·s)	130	175	250	100
B (mPa·s)	45	50	30	100
Wt. ratio A/B	2.24	1.18	2.0	3.6
T_O (°C)	25	25-40	25	60-70
T_m (°C)	95	50-70	100	120-130
Mold Pressure (MPa)	1-5	---	0.2 - 0.4	---
Gel Time (s)	8	3-60	10-20	25
Demold Time (s)	45	30-300	50-70	75-300

(a) Dow - MM 353 product literature 1986
(b) Mobay - Slocum et al. (1986), viscosities at 20°C
(c) Ashland - Arimax 1100 product literature 1986, Borgnaes et al. (1984).
(d) Shell - Farris, et al. (1983).

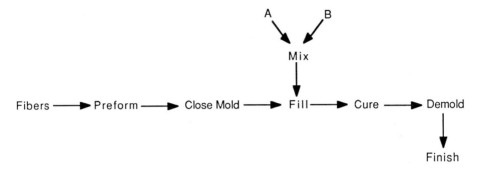

Figure 8.8   Unit operations for the structural RIM process.

8.3 Processing Mat Reinforced RIM

We can use the unit operations shown in Figure 8.8 to focus on some of the special processing problems with mat reinforced RIM.

Fibers - Glass have been the primary fibers used in S-RIM, however, graphite (Wilkinson and Eckler, 1985) and nylon (Cordova et al. 1987) have been used. As indicated in section 8.1 glass strands are actually bundles of hundreds of filaments (about 10 μm dia.). To get maximum properties each filament should be bonded to the polymer matrix. Thus the reactive mixture must wet the fibers. Wetting is discussed further below but we note here that the adhesive which holds the fibers in bundles should dissolve quickly into the reactants. Then the resin must bond to the fiber surface. Coupling agents have been developed to improve the bonding of epoxies and styrenic resins to glass but less work has been done with isocyanurates and urethanes. It is an open question whether good coupling can be achieved in the short reaction times desired with S-RIM.

The main advantage of S-RIM, its high concentration of continuous fiber reinforcement, is also a disadvantage. It is difficult to get a smooth part surface with the many long, entangled fiber bundles. If several thin mats (called "veils") made of single fibers rather than strands are placed at the mold surface, appearance improves considerably. However, it does not yet seem possible to produce high gloss surfaces (as is achievable with RIM or even R-RIM) on S-RIM parts. Thus S-RIM applications are mainly for non-appearance parts. The same problem of rough surfaces occurs in the sheet molding process. When high gloss is desired a second step is added to the process. After the initial compression the mold is reopened slightly and a reactive liquid is injected on the "appearance" surface. This is process is called "in-mold coating" (Ditto, 1978).

Preform - The simplest preform is just a stack of several random mats of either continuous or chopped glass strands placed in the mold cavity. This is adequate for large, flat pieces. But even these mats must be cut to fit very closely to the edge of the cavity. Otherwise there will be channelling of flow into these areas which can lead to poor mold filling patterns. The edges of the part will be resin rich and thus can be brittle. Resin rich areas also occur when the fiber mat is bent around a corner. The mat typically moves to the inside of a radius so there will be excess resin on the outside.

One advantage of S-RIM is that reinforcement can be placed where is needed. The preform can be made from a combination of woven mats near the surface and random mats in the center. Even a foam board can be used in the core to optimize strength, weight and cost. Uniaxial plies can be laid up at 90° to each other as is done with high strength laminates. The main drawback to these specialized preforms is their cost. Large volumes are needed to justify extensive automation. The lowest cost approach is to die cut random mats for use in simple, relatively flat S-RIM parts. For more complex shapes the simpliest manufacturing is done by spraying chopped fiber strands onto a male form. The strands, coated with a contact adhesive, stick to each other to make the preform.

A major barrier to wider structural RIM production is the step of placing the preform into the mold cavity. This must be automated to make large scale production of parts by S-RIM competitive. Development of cost effective methods for handling large, fragile fiber structures is perhaps the greatest need for S-RIM.

Close Mold - The preform thickness is usually several times greater than the cavity thickness. Thus the press must compress the preform as it closes. This compression results in high frictional forces between the preform and the mold surface and also between the fibers. They help to hold the preform in place during the filling step. Gauvin and Chibani (1988) report the pressure (press force/mat area) needed to compress several types of mat. Figure 8.9 shows their results. Batch and Macosko (1988) have fit the data for random mats with a simple Hookean model

$$p = k_o (\phi_o - \phi) \qquad (8.2)$$

where k_o depends on the fiber modulus and distance between fiber-fiber contacts; $1 - \phi_o$ is the initial volume fraction of fibers between the press plattens under very low pressure. The comparison between Equation 8.2 and the data for three random mats in Figure 8.9 is quite good.

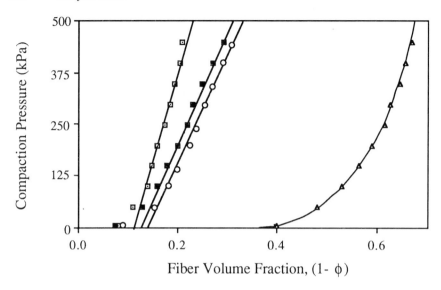

Figure 8.9   Stress needed to compress various glass fiber mats.
  □ - N-758, Nico Fiber; $1 - \phi_o = 0.11$; $h_o = 2.2$mm (thickness)
  ○ - OCF860, Owens Corning; $1 - \phi_o = 0.125$; $h_o = 2.4$mm
  △ - U 850, Vetrotex St. Gobain; $1 - \phi_o = 0.14$; $h_o = 2.6$mm
  ▲ - WR24, FIberglass Canada; $1 - \phi_o = 0.35$; $h_o = 0.8$mm
  (replotted from Batch and Macosko, 1988)

Pressure to compress a woven glass mat is also shown in Figure 8.9. The initial fiber volume is much higher and the compression is very non linear, rising sharply at higher volume fraction. Gutowski and coworkers (1986) have observed similar behavior for aligned fiber bundles. Batch and Macosko (1988) have modeled this behavior with a variable "fiber contact" approach.

Mixing - The same type of impingement mixing heads described in Chapter 4 are also used to mix the reactants before they flow into the fiber filled mold. Flow around the fibers will improve mixing and thus the critical Reynolds number for S-RIM is in general lower. However, the last material to enter the mold, just at the gate, will not flow around fibers. Mixing quality of this material can be improved by placing fibers in the runner.

Mold Filling - Filling is the crucial step of the process. Resin must quickly fill the heated mold and wet all the individual fibers before much reaction occurs. If large voids are left at the end of filling the part is ruined. If the fibers are not completely wet with resin the strength of the composite part will be reduced. Thus we need to analyze flow on two scales: distribution of liquid throughout the entire cavity around the fiber bundles and penetration into the bundles.

Wetting - Let's look at the wetting of individual fibers first. Flow into the bundles must be by capillary action. The driving pressure is created by surface tension. To roughly estimate this we can assume that the spaces between the parallel fiber bundles are cylindrical channels and use the capillary equation

$$\Delta p \; = \; \left(\frac{2\gamma \cos \theta}{R} \right) \tag{8.3}$$

where γ is the resin surface tension, θ is the resin-glass contact angle and R an effective radius of the channels. R will be on the order of the fiber radius. The penetration distance into a capillary tube under constant pressure can be derived from the Poiseuille equation and a mass balance (e.g. Tadmor and Gogos, 1979, pg. 587).

$$L \; = \; \frac{R}{2} \left(\frac{\Delta p}{2\mu} \right)^{1/2} t^{1/2} \tag{8.4}$$

We can combine Equations 8.3 and 8.4 to estimate the distance resin can travel into a fiber bundle by capillary action. If the fiber radius is 10μm, resin viscosity 1 poise, surface tension 30 dyne/cm^2 and contact angle 30°, we find that the resin will penetrate about 600μm in the first 1s. This result is in agreement with estimates by Farris et al. (1983). As yet there are not direct experimental tests of these penetration distances. However, microscopic examinations of fiber bundles after curing indicate complete penetration of the fibers (Gonzalez and Macosko, 1983). Since the diameter of the fiber bundles is on the order of 200μm these results are in agreement with our rough calculations. Thus it appears that if the bundles can be well covered with resin by the bulk flow then fiber wet out will be quick and complete.

Permeability - From the above discussion the major problem in filling is resin distribution into the cavity. Since the cavity is filled with many small fibers it can be treated as a porous media. Flow through porous media can actually be modelled more simply than an open cavity. The pores act like many tiny capillary tubes. Thus the fluid moves like a plug into the mold and there is no fountain flow (see Fig. 5.3). The average velocity (based on the empty channel) in any direction through the fiber bed is proportional to the pressure gradient and the permeability, K, in that direction

$$v_x = \frac{K_x}{\mu} \frac{dp}{dx} \tag{8.5}$$

where μ is the Newtonian viscosity.

Equation 8.5 is known as Darcy's law. It can be applied directly to filling of an end-gated rectangular mold. The velocity v_x is just the flow rate per empty cross sectional area of the mold

$$v_x = \frac{Q}{WH} = \frac{K_x}{\mu} \frac{\Delta p}{L} \tag{8.6}$$

This relation can be used to evaluate K_x by measuring Δp and Q of an unreactive liquid through a rectangular channel filled with fibers. Figure 8.10 shows results from Gonzalez (1983) for the flow of water through a chopped strand fiberglass mat (type AKM, PPG Industries). The permeability is inversely proportional to the slope. We see that K_x is a strong function of the porosity or resin volume fraction, ϕ, of the mats.

The dependence of permeability on ϕ can be fit with Kozeny's equation which is usually written

$$K_x = \frac{d^2 \phi^3}{C (1 - \phi)^2} \tag{8.7}$$

where d is the diameter of the particles or the fibers which make up the porous bed. The constant C has been found to be 150-200 for beds of packed spheres. For fiber mats the choice of d is not easy because the strands are actually bundles of fibers which may start to open up during flow. The strands are also typically not round but more like flat or elipsoidal ribbons. Gonzalez (1983) estimated his strand diameter at 54μm and compared the results of Figure 8.10 and similar graphs for radial flow to Equation 8.7 The fit shown in Figure 8.11 is good with C = 180. The fact that both rectilinear and radial flow give the same permeability means that these mats are isotropic in the mold plane, $K_x = K_y$.

Martin and Son (1986) report values of K_x similar to Gonzalez for corn syrup flowing through a continuous strand, random mat (OCF M8605, 0.045g/cm², Owens Corning). They find that the quantity d^2/C in Equation 8.7 is: $275 \pm 75\mu m^2$ for $\phi = 0.645 - 0.824$. They did observe some increase in d^2/C as pressure was increased from 0.3 - 1.4 bar. This may be due to opening up of the strands at higher flow rate.

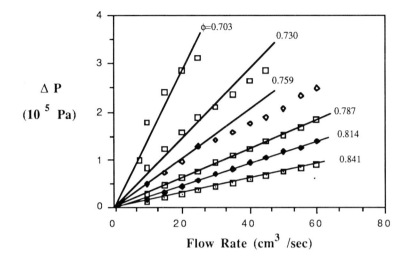

Figure 8.10 Pressure drop vs. water flow rate through a rectangular channel (540 x 100 x 6.8mm) filled with different volume fractions, 1 - ϕ, of chopped strand fiberglass mat. The slope of these lines is proportional to K_x, the permeability. At higher pressure the relation is non linear, perhaps due to deformation of the mat. (plotted from Gonzalez, 1983).

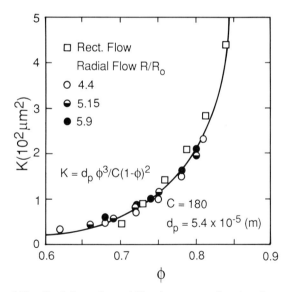

Figure 8.11 Permeability, K, of chopped strand fiberglass mat as a function of porosity. Squares are from Figure 8.10 using Equation 8.6. The circles are from similar plots for radial flow. Solid line is best fit to Equation 8.7 (adapted from Gonzalez, 1983 and Gonzalez-Romero and Macosko, 1986).

Gauvin and coworkers (1987) report data on the same mat using a slightly shear thinning polyester resin. They studied higher porosities and find a strong dependence of K on ϕ; K increases from 5 to 70 x 10^3 μm^2 as ϕ increases from 0.87 to 0.93. These porosities correspond to 6 to 3 layers of the OCF 8605 mat compressed into a 8mm cavity. Porosities above 0.9 are probably too high for S-RIM; the mats would be displaced by the flow. For a very similar density mat (N-758, Figure 8.9) over the same porosity range Gauvin et al. measured about 1/8 the permeability. This must be due to a smaller effective strand diameter. Gauvin et al. reported essentially zero permeability for a woven glass mat (WR 24). They discuss how to estimate K for a stack of mats with different permeability.

All the permeability values reported above were measured in liquid filled mats. The pressure to push liquid into a dry mat may be greater due to air bubbles trapped in between the fibers. Williams et al. (1974) found permeability about three times lower for flow into dry beds of aligned fibers with $\phi = 0.49$. Carleton et al. (1985) report they needed to evacuate a well sealed S-RIM mold to get complete filling. However, by venting through some glass clamped into the mold edge they could operate without vacuum. Slocum et al. (1986) also were able to mold without vacuum. Vented rather than evacuated molds appear to be the norm for S-RIM production.

Mold Filling - With the permeability it is possible to predict the time to fill a mold under constant pressure. For a rectangular mold from Equation 8.6 with $v_x = L/t$

$$t = \frac{\mu L^2}{K_x \Delta p} \tag{8.8}$$

For a disk shaped mold with an inlet tube of radius R_0 and part diameter of 2R

$$t = \frac{\mu}{K_r \Delta p} \left\{ R^2 \ln \frac{R}{R_0} - \frac{R^2 - R_0^2}{2} \right\} \tag{8.9}$$

However, most S-RIM machines fill under constant flow rate. Thus pressure rises until a maximum is reached. For the rectangular mold substituting $L = Qt / WH$ for Equation 8.8 gives

$$\Delta p = \frac{\mu t}{K_x} \left(\frac{Q}{WH} \right)^2 \tag{8.10}$$

For radial flow

$$\Delta p = \frac{\mu Q}{4\pi K_r H} \ln \left\{ \frac{Q t}{\pi H \phi R_0^2} + 1 \right\} \tag{8.11}$$

These equations can be compared to pressure measurements reported by Carleton et al. (1987). The mold was a rectangular plaque 914 x 406 x 3.2 mm with a center gate about 12 mm diameter. It contained only two mats (OCF 8610, 0.060 g/cm^2) and thus $\phi \cong$ 0.86. Figure 8.12 shows readings from a pressure transducer mounted 5 cm from the gate for different filling times, i.e. more mold packing. The pressure starts below gauge because the mold is evacuated. It rises rapidly due to the radial flow and then linearly after about 0.5 s as the flow becomes axial.

To compare these experiments to Equations 8.10 and 8.11 there is a problem determining the proper viscosity. Typically the mold and mats are considerably hotter than the incoming resin. Near the gate resin flow will cool down the mat but far away the resin will be close to mold temperature. Figure 8.13 shows the temperature profiles for the 2.15s shot in Figure 8.12. Note that temperature dropped from 115°C to 38°C at the pressure transducer, 51 mm from the gate, but is 60°C at 330 mm. Gonzalez-Romero and Macosko (1985) have solved for the heat transfer from the mat during filling but we can get a rough estimate of the pressure by just using an average viscosity for each flow region. For the unreacted MM 373 resin mixture μ = 33 mPa·s at 38°C (radial flow region) and about 10mPa·s at 60°C (axial flow). For the permeability we use $K_x = K_r = 3,300\ \mu m^2$ from the data of Gauvin et al. (1987) for OCF-8605 mat at $\phi = 0.86$. This value of K compares well to 3000 μm^2 calculated from Equation 8.7 with d = 130 μm, the value reported by Carleton et al., 1987. With these viscosity and permeability values we can fit the filling pressures curve in Figure 8.12 reasonably well using Equation 8.11 followed by Equation 8.10 with Q = 293 cm^3/s assuming half the flow goes to each side of the mold after 0.3 s.

At higher glass content pressure to fill the mold can exceed machine capacity. For example at $\phi = 0.6$ (five OCF 8610 mats in a 3.2 mm thick cavity) Carleton et al. (1985) report filling pressures exceeded 50 bar for the same rectangular mold used in Figure 8.12. This pressure combined with the pressure needed for impingement mixing caused the RIM machine to exceed its safety limit of about 19 MPa.

Another problem which can occur during filling is mat slip. High velocities around the fibers can cause mats to slide or tear apart especially at low glass content. Slocum et al.(1986) recommend not to exceed 500 cm^3/s. Mat tearing is caused by the drag forces

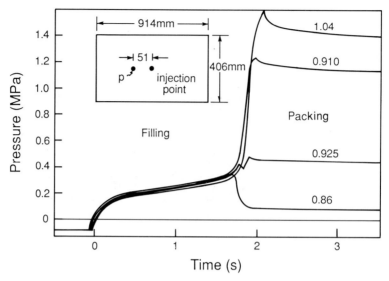

Figure 8.12 Pressure vs time during filling for increasing shot volume, expressed as fraction of total filling. Flow rate 586 cm³/s into a center-gated rectangular mold (914 x 406 x 3.2 mm) and
containing two random glass mats, φ = 0.86. Pressure measured near gate at R_0 = 51 mm.
Resin: Dow's isocyanurate Spectrim MM 373 (replotted from Carleton et al., 1987).

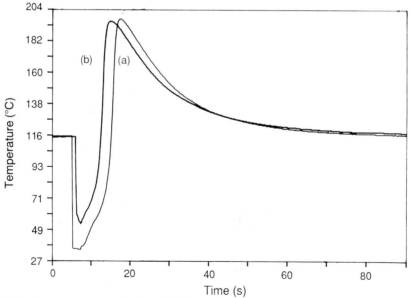

Figure 8.13 Temperature vs time during mold filling and curing, same conditions as Figure 8.12. Thermocouples are located in the center of the mats at 5 (a) and 33 (b) cm from the injection point
(replotted from Carleton et al., 1987).

generated by resin flowing over the fibers. This is counter balanced by friction of the fibers against themselves and the mold wall.

The drag force on a single cylinder is (Batchelor, 1967)

$$F_D = 4\pi C \mu v \qquad\qquad (8.12)$$

where $C^{-1} = \log (7.4/Re)$ and v is the local fluid velocity over the fibers. The resisting friction between the fibers should depend on factors similar to those which lead to Equation 8.2 for the stress to close a press. For a given volume fraction of fibers the friction force should be constant. Thus for a particular mold, mat tearing will start if F_D increases beyond a critical value. This can happen if viscosity or velocity increase.

Further theoretical and experimental work are needed to better understand the limitations imposed by mat strength. However, we can draw a qualitative moldability diagram for S-RIM filling similar to that for RIM shown in Chapter 5, Figure 5.19. A major difference is that the mold wall temperature, T_w, is more important during filling in S-RIM. This is because resin flows through the porous mat as a plug. Liquid at the front quickly reaches the temperature of the mat. S-RIM is typically carried out as a heat activated process; T_0 $\ll T_w$. Thus the maximum time to fill a given mold is controlled by the gel time at T_w

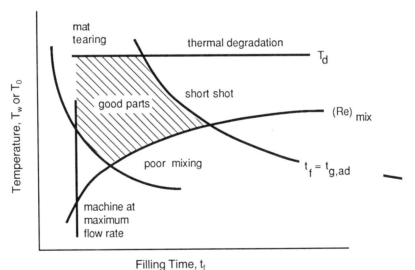

Figure 8.14   Moldability diagram for filling in S-RIM. Mixing orifices, volume fraction of glass, and resin chemistry are fixed. Degradation and premature gelation are controlled by mold temperatrue, T_w while mat tearing and poor mixing are governed by viscosity of the resin at its initial temperature, $\mu(T_0)$.

$$t_f < t_g (T_w)$$
(8.13)

Degradation is also controlled by mold temperature. However, as in RIM, initial resin temperature determines viscosity which controls mixing quality. Most S-RIM parts are center gated. Thus they initially fill via radial flow. The highest velocity is at the gate where viscosity is that of the unreacted resin mixture at T_0. Thus F_D, the initial drag force for mat tearing, can be controlled by the flow rate ($t_f \propto 1/Q$) and the initial temperature through $\mu(T_0)$. A schematic moldability diagram for S-RIM is shown in Figure 8.14.

One solution to difficult filling and for filling of large parts is multiple gating as discussed by Martin and Son (1986). An obvious problem with multiple gates in non evacuated molds is the trapping of air as two flow fronts approach each other.

Curing - The curing step in S-RIM is similar to RIM as discussed in Chapter 6. The major difference is the temperature distribution at the end of filling due to heat transfer from the mat. This is illustrated in Figure 8.13. 33 cm from the gate at the end of filling the isocyanurate is much hotter and polymerizes more quickly than close to the gate.

Gonzalez-Romero and Macosko (1985) have solved the problem of heat transfer during transient flow through a porous bed. The solution is analytical for the adiabatic case with no reaction, i.e. no heat transfer from the mold walls, only exchange between the porous solid and the flowing liquid. They also neglected temperature gradients inside each fiber. They found experimentally the heat transfer coefficient between a petroleum oil and glass fibers was linear with velocity

$$h = \lambda v$$
(8.13)

with $\lambda = 44$ J/m^{3}K.

For radial flow v is decreasing but Gonzalez (1983) found that an average heat transfer coefficient could fit his experimental data for oil and for a styrene dimethacrylate monomer mixture. The left side of Figure 8.15 shows some of his results at three different radial positions. We see that as the cold resin flows over the hot mats temperature drops essentially to the initial resin value. However, the flow front picks up all the heat from the glass mat. Thus at the end of filling the temperature at the mold perimeter is nearly that of the initial mold with a steep gradient back to the gate. Gonzalez' measurements showed that at any radial position temperature was nearly constant with thickness rising to T_w only at z/H > 0.7. This is a test of the adiabatic filling assumption. This will be even more valid at higher flow rates, i.e. shorter fill times.

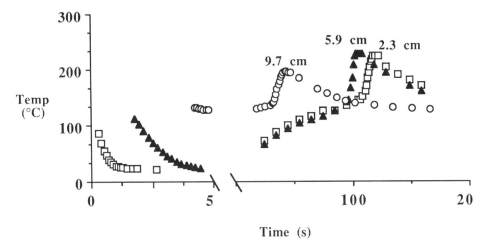

Figure 8.15   Temperature profiles during filling and curing of a radial mold (R = 9.93cm, H = 0.52 cm, R_0 = 0.5 cm) containing random glass mats, ϕ = 0.764. Resin is a free radically cured styrene dimethacrylate mixture. Q = 27 cm^3/s (replotted from Gonzalez, 1983).

The radial temperature profile at the end of filling is the starting point for curing. Using these initial values and neglecting radial and axial heat transfer, temperature and conversion are calculated at each position using the one dimensional equations given in Chapter 6. The only difference is the presence of the fibers. They modify the Damköhler number, Da, reducing the reactant concentration and thus the maximum temperature rise and modify the thermal diffusivity. Using Equations 6.4 and 6.5 with accurate kinetic data curing temperatures can be predicted as illustrated in the right side of Figure 8.15. Conversion will follow the temperature profiles. The resin will gel first at the mold perimeter. A gelling front will move in toward the gate. If there is considerable shrinkage during polymerization there will be a sink mark around the gate.

Figure 8.16 shows the influence of resin content on the curing exotherms. For this fast polymerizing system the maximum exotherm decreases almost linearly with increased glass volume fraction, 1 - ϕ. Curing is also a bit faster due to more heat transfer from the glass during filling.

A heat transfer model combined with an accurate kinetic equation for the polymerization can provide useful insight into process and resin variation. For example Figure 8.17 illustrates the large effect of initiator level on curing of a free radical polymerization. At high initiator concentration the maximum exotherm is reduced because reaction starts before the resin has reached mold temperature.

Packing - Although S-RIM is a low pressure process some packing pressure at the end of filling is necessary to reduce the amount of trapped gas in the part. If we look back to Figure 8.12 we see that pressure at the end of filling was increased by increasing the shot volume (actually shot time). Carleton et al. (1987) measured density profiles in these parts and found that it was necessary to pack the mold with about 1.2 MPa pressure to get uniform density. The void content in this part was about 1.5%. Higher pressures opened up his clamp and thus did not result in further increase in density. There don't appear to be any studies to determine whether this void content leads to reduced mechanical properties such as fracture or fatigue. If this is the case it would be valuable to study the influence of packing pressure further.

Demolding - Demold time in S-RIM is controlled by the same factors discussed for crosslinking RIM systems in Chapter 6. Moldability diagrams like Figure 6.24 are still valid. Minimum demold time is determined by time for the last resin to enter the mold to cure, that is near the gate. The conversion at demold can be lower than with RIM since the fiber mats greatly increase part stiffness. However, without post curing it is important to have 95% or higher conversion of reactive groups in most chemical systems. Because of the high modulus of S-RIM even at elevated temperatures (note Figure 8.7) parts can be removed from the mold hot without as many problems with warpage.

Finish - As discussed under the preform step there is often excess resin at the edges of the part and in the vents. If glass is clamped in the parting line as is usually done with RTM then considerable trimming and sanding of the edges is required.

All the process unit operations in S-RIM (and in RIM) are coupled to each other, they cannot always be changed independently. Moldability diagrams like Figure 8.14 and those for RIM in Chapters 5 and 6 can help engineers understand this coupling and more efficiently select suitable process parameters to start process optimization. Gonzalez-Romero and Macosko (1986) present a strategy for selecting S-RIM process parameters. Their logic diagram is shown in Figure 8.18. The process engineer starts with a chemical system and reinforcement level selected based on physical properties and cost needed in the final part. Figure 8.7 is one example of how ϕ might be specified based on the needed part stiffness.

From the volume of the part, ϕ and the machine capacity an initial estimate is made of flow rate. Using Equations 8.10 and 11 the fill time and maximum pressure can be estimated. With the polymerization kinetics at T_w and knowledge of the gel conversion a gel time, t_g, at the flow front can be calculated. The same kinetics can also be used combined

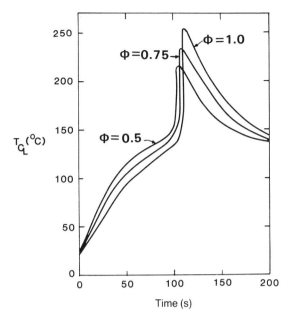

Figure 8.16 Effect of volume fraction of resin on mold exotherms for the same resin used in Figure 8.15. H = .65 cm. No heat transfer during filling, representative of the gate region (from Gonzalez, 1983).

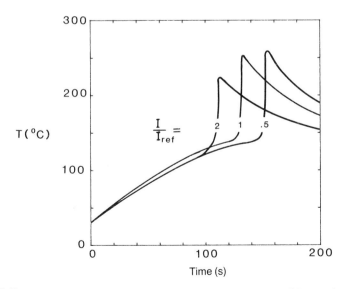

Figure 8.17 Influence of initiator level on mold exotherms. Same conditions and resin as Figure 8.16 with ϕ = 0.8 (from Gonzalez, 1983).

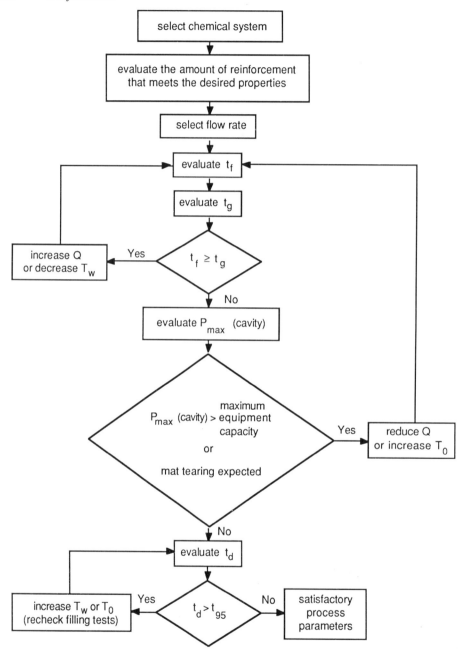

Figure 8.18   A strategy for selecting process parameters for S-RIM. It is assumed that mixing and degradation criteria are always satisfied (Gonzalez-Romero and Macosko, 1986).

with the heat transfer analysis (Equations 6.4 and 6.5) starting at T_0 to estimate demold time.

REFERENCES

Batch, G.L.; Macosko, C.W. "A model for two-stage fiber compaction in composite processing," *SAMPE Int. Tech. Conf.* Minneapolis, Sept. 1988.

Batchelor, G.K. "An Introduction to Fluid Mechanics," Cambridge University Press, Cambridge UK, 1967.

Borgnaes, D.; Chapell, S.F.; Wilkinson, T.C. "RIMing of low viscosity crosslinkable alcohols with diisocyanates," Soc. Plast. Ind., Urethane Conf. 1984, pg. 381-386.

Carleton, P.S.; Breidenbach, D.J.; Proctor, G.C. "Pressure measurement in mat molding (MM/RIM)," Am. Soc. Mat. Adv. Comp. Conf., Detroit MI, Sept. 1987, paper 8707-012.

Carleton, P.S.; Waszeciak, D.P.; Alberino, L.M. "A RIM process for reinforced plastics," Am. Soc. Metals, Adv. Composites Conf., Dearborn MI, Dec. 1985, paper 8521-009.

Carleton, P.S. 1987, private communication.

Chisnall, B.C.; Thorpe, D. "RRIM - a novel approach using chopped fibreglass," Soc.Plast. Ind., Reinforced Plastics/Composites Conf., 1980.

Coates, P.D.; Sivakumar, A.I.; Johnson, A.F. "RRIM: rheology of glass-filled polyol slurries and polymeric isocyanate reagents," *Plast. Rubb. Process. Appl.* **1987**, *7*, 19-28.

Cordova, C.W.; Young, J.A.; Rowan, H.H. "Nylon fiber reinforcement for polyurethane composites," *Polym. Comp.* **1987**, *8*, 253-256.

Cross, M.M.; Kaye, A.; Stanford, J.L.; Stepto, R.F.T. "Rheology of polyols and polyol slurries for use in reinforced RIM," in "Reaction Injection Molding," J.E. Kresta, Ed., Am. Chem. Soc. Symp. Series 270: Washington D.C., 1985, p.97-110.

Ditto, E.D. "Mold coating of freshly molded articles," US Patent 4,076,788, **1988**. General Motors.

Farris, R.D.; de la Mare, H.E.; Overcashier, R.H.; Gottenberg, W.G. "Structural parts from epoxy RIM using preplaced reinforcement," *Polym.-Plast. Technol. Eng.* **1983**, *21*, 129-157.

Ferrarini, J.; Cohen, S. "Reinforcing fillers in RIM PUR," *Mod. Plast.* **1982**, *59 no.10*, 68-71.

Gauvin, R.; Chibani, M.; Lafontaine, P. "The modeling of pressure distribution in resin transfer molding," *J. Reinf. Plast.* **1987**, *6*, 267-274(?).

Gauvin, R.; Chibani, M. "Modelization of the clamping force and mold filling in resin transfer molding," *43rd Ann. Conf. Reinf. Plast./Comp. Inst., Soc. Plast. Ind.* **1988**, 22-C.

Gerkin, R.M.; Lawler, L.F.; Schwartz, E.G. "Reinforcement systems for high modulus reaction injection molded urethane composites," *J. Cell. Plast.* **1979**, *15*, 51-58.

Girgis, M.M.; Das, B.; Harvey, J.A. "Fiber glass reinforced reaction injection molding," in "Polyurethane Reaction Injection Molding and Fast Polymerization Reactions,"Proc.of Int. Symp. on Reaction Injection Molding, Am. Chem. Soc., Atlanta, March 1981, Plenum: New York, 1982, p.219-229.

Girgis, M.M. "Fiberglass reinforcements in RIM urethanes," in "Reaction Injection Molding," J.E. Kresta, Ed., Am. Chem. Soc. Symp. Series 270: Washington D.C., 1985, p.225-236.

Gonzalez, V.M.; Macosko, C.W. "Properties of mat reinforced reaction injection molded materials," *Polym. Comp.* **1983**, *4*, 190-195.

Gonzalez, V.M.; "Studies of reactive polymer processing with fiberglass reinforcement," PhD Thesis, Univ. Minnnesota, **1983**.

Gonzalez-Romero, V.M.; Macosko, C.W. "Adiabatic filling through packed beds in composite reaction injection molding," *Polym. Proc. Eng.* **1985**, *3*, 173-184.

Gonzalez-Romero, V.M.; Macosko, C.W. "Process parameter estimation for composite reaction injection molding and resin transfer molding," *Soc. Plast. Eng. Tech. Papers* **1986**, *31*, 1292-94.

Gutowski, T.G.; Kingery, J.; Boucher, D. "Experiments in Composites Consolidation: fiber deformation," *Soc. Plast. Eng. Tech. Papers* **1986**, *32*, 1316-1320.

Martin, G.Q.; Son, J.S. "Fluid mechanics of mold filling for fiber reinforced plastics," Am. Soc. Mat., Adv. Composites Conf., Dearborn MI, Nov. 1986.

Martinez, E.C.; Vanderhider, A. "On line paintable RIM body panels," Soc. Auto. Eng. Int. Cong. and Exposition, Detroit, Feb. 1987.

Nelson, D. "High-strength structural composites made by RIM," *Plast. Eng.* **1987**, *11*, 29-32.

Oertel, G., ed., *Polyurethane Handbook*, Hanser: Münich, 1985.

Rice, D.M.; Dominguez, R.J.G. "Impact properties of reinforced RIM fascia," *J. Cell.Plast.* **1983**, *19*, 114-120.

Slocum, G.H.; Nodelman, N.H.; Fluharty, C.E.; Schumacher, D.W. "Structural RIM: successful combination of RIM process and fiber reinforcement," Am. Soc. Mat., Adv. Composites Conf., Detroit, Nov. 1986.

Tadmor, Z.; Gogos, C.G. "Principles of Polymer Processing," Wiley, New York, 1979.

Taylor, R.P., Mobay Chemical Corp., private communication, 1988.

Tucker, C.L.; Suh, N.P. "Mixing for reaction injection molding. II. Impingement mixing of fiber suspensions," *Polym. Eng. Sci.* **1980**, *20*, 887-898.

Wilkinson, T.C.; Eckler, J.H. "Advances in structural RIM," Soc. Plast, Ind., Urethane Conf., 1985, pg 59-62.

Williams, J.G.; Morris, C.E.M.; Ennis, B.C. "Liquid flow through aligned fiber beds," *Polym. Eng.Sci.* **1974**, *14*, 413-419.

SUBJECT INDEX